AF553035

Ecology and Human Development

NIPA® GENX ELECTRONIC RESOURCES & SOLUTIONS P. LTD.
New Delhi-110 034

About the Authors

Dr. Mouneshwari Ramanand Kammar is a distinguished academician and extension professional, presently serving as Professor (Human Development and Family Studies) and Technical Officer to the Director of Extension at the University of Agricultural Sciences, Dharwad. She holds a B.H.Sc. in Home Science (1994), M.H.Sc. in Human Development (1997), and a Ph.D. in Human Development (2009) from SNDT Women's University, Mumbai. Additionally, she has pursued an MBA in Human Resource Development (2009), PG Diploma in Nutrition and Dietetics (2012), and PG Diploma in Agricultural Extension Management (2013).With over two decades of service, Dr. Kammar has held significant roles in Krishi Vigyan Kendras at Tukkanatti, Raichur, and Bagalkote, later advancing as Senior Scientist and Head, Associate Professor, and Professor. Her career reflects a strong commitment to rural development, women empowerment, and capacity building. She has received several prestigious honors, including the Choudhary Devilal Award (2003), Scientist of the Year Award (2011), and Pandit Deen Dayal Upadhyay Krishi Vigyan Rastriya Protshahan Puraskar (2019). A prolific writer, she has authored 60+ research papers, 15 books, and over 150 articles, contributing significantly to human development and agricultural extension.

Dr. Vinutha U. Muktamath is a distinguished researcher in Human Development and Family Studies. She earned her B.H.Sc. and M.H.Sc. in Human Development from the University of Agricultural Sciences (UAS), Dharwad, graduating with distinction, and completed her Ph.D. in the same discipline in 2017. Additionally, she holds a Postgraduate Diploma in Agricultural Extension Management from MANAGE, Hyderabad. She has over two decades of teaching, research, and extension experience, having served as Scientist and Assistant Professor at UAS Dharwad and earlier as Agricultural Officer with the Karnataka State Department of Agriculture. Her academic contributions include more than 25 research papers in reputed national and international journals and over 12 authored and co-authored books/chapters in English and Kannada, covering topics such as maternal and child health, breastfeeding practices, elderly well-being, and community development.Dr. Muktamath has successfully completed several sponsored research projects on nutrition, postnatal care, and elderly empowerment. For her outstanding contributions, she has received multiple awards and recognitions in teaching, research, and extension services.A committed academician, she continues to contribute to advancing knowledge and practice in family health, child development, and sustainable community welfare.

Dr. Prema Patil is a professor in the Department of Human Development and Family Studies, College of Community Science, University of Agricultural

Sciences, Dharwad, Karnataka, India and has a work experience of 25 years. She has completed her PG Diploma in Psychological Counseling, Rural Development and Family Life Education. She has received gold medal in her Master's and PhD degree. She was conferred the Best Extension Worker award by University of Agricultural Sciences, Dharwad in the year 2020 and three awards by professional societies. She was the Principal Investigator of a project funded by Department of Science and Technology, New Delhi and four Staff Research projects funded by UAS Dharwad, Karnataka. She has completed 21 trainings from reputed institutions. She has published 37 research articles, 35 popular articles and attended 14 conferences at national and international levels.

Ms. Sunkara Hanna is currently pursuing her Ph.D. (1st year) in Human Development and Family Studies at the University of Agricultural Sciences, Dharwad. In 2024, she worked as an Assistant Professor in the Department of Human Development at the Avinashilingam Institute of Home Science and Higher Education for Women, Coimbatore. She holds a B.H.Sc in Community Science and a Master's in Human Development and Family Studies, securing top national ranks in ICAR JRF and SRF examinations, and has also qualified for UGC NET JRF. Her research focuses on cognitive function and health among the elderly, with several publications in reputed journals, books, and popular articles. Passionate about teaching and research, she specializes in child development, family relations, and community well-being, and actively contributes to seminars and conferences at both national and international levels.

Ms. Madhumati K Mantur, Ph.D scholar at the University of Agricultural Sciences, Dharwad specialization in Human Development and Family Studies and also holds M.H.Sc and B.Sc. in Home Science, along with PG Diplomas in Computer Applications and Information Technology. She has served as a Teaching Associate at the College of Community Science, ANGRAU, Guntur and as a Lecturer at Murarji Desai Residential College, Maragur with responsibilities in teaching, mentoring and academic supervision. She has cleared NET and KSET and received recognitions such as the International Excellent Education Award, Young Professional Award and Excellence in Extension. Her research and publications focus on sibling relationships, socio-emotional behavior, family dynamics, digital exposure and consumer preferences, alongside the development of practical manuals in family studies and early childhood education. She has actively contributed to extension and outreach programs on child welfare, health, pedagogy, disability inclusion and women's empowerment and has shared her work through conferences, workshops, faculty development programs and radio talks.

Ecology and Human Development

Mouneshwari R Kammar
Vinutha U Muktamath
Prema Patil
Sunkara Hanna
Madhumati K Mantur

Department of Human Development and Family Studies
College of Community Sciences
University of Agricultural Sciences
Dharwad, Karnataka, India

NIPA® GENX ELECTRONIC RESOURCES & SOLUTIONS P. LTD.
New Delhi-110 034

NIPA® GENX ELECTRONIC RESOURCES & SOLUTIONS P. LTD.

101,103, Vikas Surya Plaza, CU Block
L.S.C. Market, Pitam Pura, New Delhi-110 034
Ph : +91-11-43860225, Mob.: +91 9717133558, 9540816132
E-mail: newindiapublishingagency@gmail.com
Website: www.nipaersources.com

Print ISBN: 978-93-58873-46-7
ebook ISBN: 978-93-58875-58-4

Composed and Designed by NIPA®.

Preface

Human development is a dynamic and lifelong process influenced by a wide range of biological, psychological, social, cultural, and environmental factors. In recent decades, the study of ecology and its relationship with human behavior and development has gained significant importance due to rapid urbanization, technological advancements, environmental changes, and shifting socio-cultural structures. Understanding these ecological influences is essential for comprehending how individuals grow, adapt, and function within their environments across the lifespan.

The present book, Ecology and Human Development, has been designed to provide a comprehensive understanding of the interrelationship between ecological systems and human development. The book explores how physical, social, cultural, technological, and policy environments shape developmental processes from childhood to old age. It integrates classical ecological theories with contemporary perspectives, thereby offering readers a holistic framework for understanding human growth and behavior in changing ecological contexts.

The contents of this book cover a broad spectrum of topics, including ecological modeling in human development, physical ecology, socio-cultural ecology, developmental ecology, systems and policies, and the impact of media and technology on human life. Special emphasis has been given to Bronfenbrenner's Bio-ecological Model and the Process–Person–Context–Time (PPCT) framework, which provide strong theoretical foundations for understanding developmental interactions within multiple environmental systems. The book also discusses the influence of climate, demographics, neighborhood, family systems, social institutions, cultural practices, disasters, technology, and public policies on human development.

Efforts have been made to present the subject matter in a simple, systematic, and reader-friendly manner. Relevant examples, conceptual explanations, and interdisciplinary perspectives have been incorporated to make the content useful for students, teachers, researchers, extension professionals, and practitioners in the fields of Human Development, Family Studies, Community Science, Psychology, Sociology, Education, and allied disciplines.

This book is expected to serve as a valuable academic resource for undergraduate and postgraduate students, competitive examination aspirants, and professionals interested in understanding ecological perspectives of human development. It is our sincere hope that the readers will gain meaningful insights into the complex interactions between humans and their environments and appreciate the importance of sustainable and development-oriented ecological practices.

We express our heartfelt gratitude to all our teachers, colleagues, students, researchers, and family members whose encouragement and support inspired us during the preparation of this book. We are also thankful to the publisher for their cooperation and support in bringing out this publication successfully.

Constructive suggestions from readers and academicians for further improvement of the book will always be welcomed.

Authors

Contents

1

Ecological Modeling in Human Development

1.1 Human ecology - its concept and relevance in understanding human development. Life course development through the ecological perspective

What is human ecology? Human ecology is an approach to the study of human behavior marked by two commitments. First, human ecologists think that humans should be studied living systems operating in complex environments. The human sciences are balkanized into several social science, humanistic, and human biological disciplines. Ecologists are used to thinking that systemic nature of individual organisms and populations of organisms mean that we typically have to understand how diverse parts of the system operate together to produce behavior. The traditional human science disciplines take people apart; human ecologists endeavor to put us back together. Breaking complex problems down to operationally tractable parts is a great strategy, but only so long as some are committed to putting them back together in the end! Second, human ecologists think that humans are subject to very similar ecological and evolutionary processes as any other species.

Human ecology is a synthetic cross-discipline. The ecological approach is attractive to many scientists because it provides a broad view, congenial to the synthesis of the contributions of the many disciplines that are required to understand human behavior. Much as the whole field of biology is united by the neo-Darwinian synthetic theory of evolution, so human ecologists seek to develop an "**umbrella theory**" to unite and make sense out of the specialized contributions of the narrower disciplines. Many disciplines contribute to human ecology because our behavior is complex and diverse. Individuals who have taken the ecological/evolutionary approach to humans include biologists, anthropologists, psychologists, sociologists, demographers, historians, geographers, geophysicists, and economists

We interact with each other and with our physical environments. We are biological creatures who depend upon the living landscape to sustain us. Plants and animals are affected by our actions, and our existence is impacted by plants and animals. We exist within complex sets of interactions – that is, we live in an ecological world. Learning to perceive the world as a never-ending system of interactions – that is, to think about our surroundings and our relationships with our environments and each other ecologically – is challenging. Such thinking forces us to rethink our views of economics, politics, and business. It suggests different ways to plan and design. In economics, for example, an ecological view suggests a much more complex set of relationships than supply and demand: supply of what and where from and at what cost, not only in dollars but to other species and other generations. Ecological understanding can also confront our values and religious beliefs, although most faiths address human connections to the natural world and stewardship responsibilities for future generations. **Ecology** is, by definition, the reciprocal relationship among all organisms and their **biological and physical environments.** People are organisms. As a result, we can ask, is the use of human as a modifier to ecology necessary? Human with ecology helps reinforce the reality of our place in environments. Human ecology seeks to understand the multiple interrelationships between the human species and our environments. Human ecology is **broader than biology**, but also is grounded in biological concepts. The **trans-disciplinary** field can be defined as the study of the **complex and varied system of interactions between people and their environments**.

Borrowing Concepts from Biology

The basic rationale for human ecology is that concepts and methods shared with the biological sciences ought to be useful to understand human behavior. Our behavior is taken to be just a special case of general ecological processes (as any particular species is a special case). This idea has a long history—in demography, for example. Malthus' pioneering ideas about human population explosions played a large role in Darwin's thinking about all populations. Darwin's ideas about natural selection in turn have had a large influence on how we think about humans. As Foley's title in the epigraph indicates, humans may be a peculiar beast, but then so is every other species. We agree with Foley that humans can't stand in some splendid isolation from the rest of nature

Malthusian principle states that populations grow geometrically while resources grow only arithmetically. Simply stated: **Population grows much more quickly than the food supply**. Malthus believed that the inability of available resources to keep pace with ever increasing population size ultimately results in a continuing struggle for survival by the lower economic classes. His Essay on the Principles of Population describes the outcome of mankind's struggle to obtain increasingly limited resources as a life filled with misery and vice.

It was this concept of a struggle for survival within large populations that was adopted by evolutionary biologists, forever changing evolutionary thought.

The impact of Malthus's Essay on the Principles of Population on Charles Darwin as he sought the mechanism for evolution has never been understated. Darwin himself recorded in his 1876 autobiography the following:

"In October 1838, that is, fifteen months after I had begun my systematic inquiry, I happened to read for amusement Malthus on Population, and being well prepared to appreciate the struggle for existence which everywhere goes on from long-continued observation of the habits of animals and plants, it at once struck me that under these circumstances favourable variations would tend to be preserved, and unfavourable ones to be destroyed. The results of this would be the formation of a new species. Here, then I had at last got a theory by which to work."

Malthus's work caused Darwin to refocus on a bigger picture. While Darwin's predecessor, Jean Baptiste de Lamarck (1744-1829), and contemporaries such as Jean Agassiz (1807-1873) and Richard Owen (1804-1892) focused on individual organisms and the belief that something was driving the evolution of organisms in a direction toward perfection independent of environmental influence, Darwin dismissed this popular theory and continued his quest for another explanation. Upon reading Essay on the Principles of Population, both Darwin and Alfred Russel Wallace independently adapted Malthus's "**struggle for existence**" principle and applied it to plant and animal species, thereby arriving at the theory of **natural selection**. Darwin's theory of natural selection stretches Malthus's principle beyond the boundaries of the human population and political economy. Darwinian theory of natural selection made the connection between organisms and their environments stronger than they had ever been.

What Will We Use From Biology?

The basic common core of ecology and evolution is sometimes called **population biology**. Human ecology borrows a complex of ideas from population biologists. The most basic of these are the ideas of population growth and regulation in a single population, as developed by demographers. If we add heritable variation, such as genetic variation to the population, then different types will compete. Some will survive and reproduce better than others, and the more fit types will replace the less fit. This is Darwin's idea of natural selection. Since the effect of natural selection depends upon the environment--a variant that fails in one environment may succeed in another-natural selection tends to produce diversity. We often say that it adapts organisms to the environment that they live in. Then, we need to think about individuals of a population interacting with each other as well as the outside environment. The evolution and ecology of social interactions is often called **socio biology**.

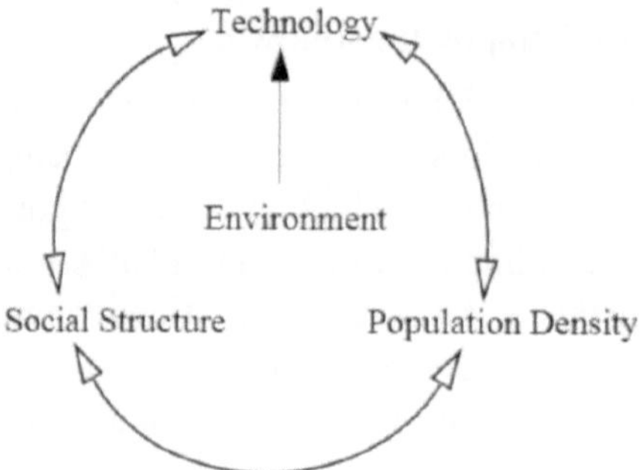

The classic "culture core" (Diagrammatic representation of culture core concept) model of how we're necessarily connected to the environment. To preview, people have to make a living by extracting resources from the environment. So do all organisms. Typical organisms use organic structures directly to make a living; lions kill prey with their teeth and monkeys grind hard seeds with their teeth. People do a little of the same, but most of our adaptations revolve around complex traditional skills we have learned from others. Human populations have a given basic set of tools (technology), whatever their evolving cultural tradition has developed to that point. The details of the toolkit will vary adaptively in the context of the given type.

Next, we can try to use community ecology, the study of how individuals and populations interact with each other via, competition, predation, parasitism, and mutualism. For example, if competitors are too similar in their resource use, the more efficient user of resources drives out the less. This is called competitive exclusion, and is rather like natural selection at the level of species instead of genes. Ecological communities tend to be composed of specialized species, each occupying a unique ecological niche. It is often useful to think of the community of organisms plus the interacting abiotic factors on a give site as a system of interacting parts, and ecosystem. We also need to recall that the organisms have effects on environments as well as vice versa. The important early 20th Century evolutionist R.A. Fisher called this environmental deterioration. Often, when one population evolves, say cheetahs become leaner and faster, the environment deteriorates as far as prey like impala are concerned. Now, selection will start to favor faster, more alert impala too, which then deteriorates the environment for cheetah. This cyclical round of deterioration and evolutionary response is termed **co evolution**

Borrowing by Homology

We can apply ecology and evolution to humans as just another animal. This is making use of homology. **Humans really are a species of ape after all**. We share many basic anatomical features with other apes, as well as subtler things. For example, all apes have a very long **juvenile dependent period** compared to most mammals, and there is likely to be some common evolutionary ecological reason why we share this feature and common con sequences for the rest of our behavior and relations to the environment. Apes have very rudimentary abilities to learn language. Many people are studying ape protolanguage with the idea that the brain structures that permit apes to have a little language served as the basis for our much expanded capacities. Comparative evolutionary ecologists often cast a wide net in animal comparisons, under the assumption that all **ecologically similar creatures will follow a similar logic.**

Borrowing by Analogy

Some of the unique attributes of the human species are only loosely analogous to commonly studied biological phenomena. We may still wish to borrow ideas from biology because the human phenomenon is similar enough to the non-human to get a good inspiration for theory or method from the borrowing. There is actually a long history of borrowing analogies back and forth between biology and social science. "Natural selection" is derived from an analogy with plant and animal breeding--artificial selection. The ecologists' term"community" is derived from an analogy with human communities.

Several analogies have attracted attention

1. **Culture is like genes**: Humans are unusual in the degree to which we get our behavior by imitation from our parents and others. Getting ideas by imitation is somewhat analogous to genetic transmission.
2. **Human societies are like species**: In most species, all populations have the same basic adaptation. Human adaptations are much more diverse. Some populations are mainly plant eaters, others are mainly predators.
3. **Human societies are like ecological communities**: Within any one human society, there are a few to many subgroups specialized to do different things. Minimally, human societies usually specialize tasks by age and gender.
4. **Human societies are like bee, ant, and termite colonies**: Only a few species live together in large numbers and cooperate extensively. Biologists call these species **ultra-social**. Many of the more social of the social insects have hundreds of thousands of individuals each the colony, all cooperating, dividing tasks, and the like. Humans are also

counted among the ultra-social animals, although the means we use to achieve ultra-sociality are apparently quite different

5. **Humans are like peacocks, bowerbirds, song-sparrows, flowers, and tropical fish**

Testing Human Ecology: Ecology and evolutionary biology are sources of hypotheses to test.

Concept, Interpretations and Definitions

Human ecology, the study of the relationships between humans and their environments, is a field with a large scope and complex history. Human ecology is an interdisciplinary and trans-disciplinary study of the relationship between humans and their natural, social, and built environments. The philosophy and study of human ecology has a diffuse history with advancements in ecology, geography, sociology, psychology, anthropology, zoology, epidemiology, public health, and home economics, among others.

The word ecology has been derived from the Greek words **oikos** meaning habitation or **home** and **logos** meaning discourse or **study** (Richerson and McEvoy, 1976). It therefore, implies a study of the **habitations of organisms**. The term was introduced by Retter in 1868 but it was properly defined by Ernst Hackel that is why he is falsely credited for coining this term (Richerson and McEvoy, 1976, Jaiswal, 2017). Ecology may be studied with particular references to animals or plants, hence animal ecology and plant ecology. Animal ecology however cannot be properly understood without a considerable knowledge of plant ecology. Therefore, animals and plants are given equal emphasis and the term **bioecology** is often used irrespective of animals or plants under study. Now a days two main subdivisions of ecology are recognized viz. **autecology** and **synecology** (Sharma, 1996; Jaiswal, 2013).

Basic Definition

Human ecology is the study of the interactions of humans with their environments, or the study of the distribution and abundance of humans. This definition is based directly on conventional definitions of biological ecology. Ecology is usually defined as the study of interactions of organisms with their environments and each other. Thus, using the term **"human ecology" actually expresses a broad ambition to understand human behavior**

Autecology And Synecology

i) **Autecology**: **Autecology** is the study of inter-relationships of the organisms of a species to biotic or abiotic environment. It is also known as **species ecology**. Autecology deals with nutrition, growth, reproduction, development and life history of the individuals of the individuals of a species in a given environment (Berry, 1976).

ii) **Synecology**: **Synecology** is the study of inter-relationship of different groups of the organisms living such as populations, biotic communities and ecosystems and their environment, which are associated together as a unit. It can be differentiated into population ecology, community and ecosystem ecology or the ecology of population, ecology of community and ecology of habitat (Berry, 1976).

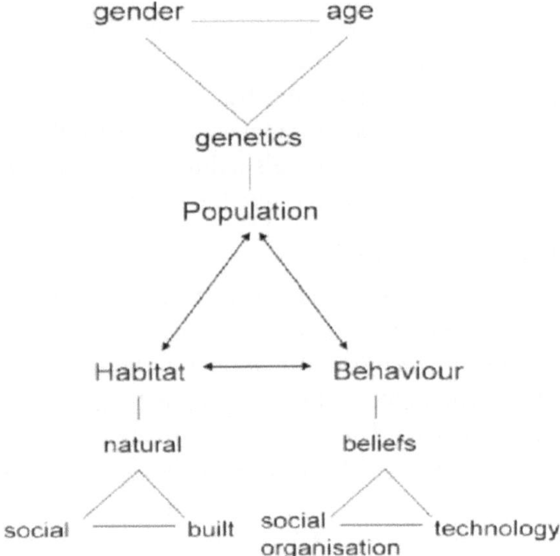

Triangle of human ecology

Ecology is the science studying how organisms interact with each other and their physical environment. These interactions are often quite complex.

Human ecology is a discipline that deals with the relationship between humans, human societies, and their natural, social and created environments. Human activity frequently disturbs living systems and affects these interactions. Ecological predictions are, of a consequence, often more general than we would like.

Human ecology is the discipline that inquires into the patterns and process of interaction of humans with their environments. Human values, wealth, life-styles, resource use, and waste, etc. must affect and be affected by the physical and biotic environments along urban-rural gradients. It arose out as scientists struggled to make sense of the impact of humans on the man-made and natural environment and the impact of environments on the social systems of humans. **The nature of these interactions is a legitimate ecological research topic and one of increasing importance.**

Human Ecology is the study of the interactions between man and nature in different cultures. Human Ecology combines the ideas and methods from several disciplines, including **anthropology, sociology, biology, economic history and archeology**. The **multidisciplinary** approach enables to comprehensively address issues of environmental justice, sustainability and **political ecology**. It studies human life and human activity in different ecosystems and different cultures in the present and in the past in order to gain a better understanding of the factors which influence the interaction between humans and their environment. The ambition to achieve a more complete view requires an integrated perspective that transcends traditional boundaries between the **humanities, social sciences, natural sciences, and technology**. Fundamental issues in human ecology are how people's cultural beliefs about the nature affect and are affected by their livelihoods and the social order.

Historical Development of Human Ecology

The earliest mention of human ecology can be found in the early 1900s among animal ecologists, who, as a result of studying population trends among plants and animals, suggested that ecological principles also applied to humans and their relationship to the natural environment. In 1859, the French Zoologist, Isodore Geoffroy St. Hilaire (Encyclopaedia Universalis; 2010) had proposed the term **ethnology** for "**The study of the relations of the organisms within the family and society in the aggregate and in the community**". At about the same time the English naturalist, **St. George Jackson Mivart** coined the term **hexicology**, which he defined in 1894 as "**devoted to the study of the relations which exist between the organisms and their environment as regards the nature of the locality they frequent, the temperatures and the amounts of light which suit them, and their relations to other organisms as enemies, rivals, or accidental and involuntary benefactors**".

Although, there is uncertainty about the original coining of the term, there is consensus that the German biologist, **Ernst Haeckel** first gave substance to this term. Haeckel, although appears to have first used the term in 1886 (Sharma, 1996) and he regarded the ecology of an organism as "….**the knowledge of the sum of the relations of organism to the surrounding outer world, to organic and inorganic conditions of existence**…." Ecology as a distinct discipline grew out of natural history early in this century as natural historians began to collect their observations into a body of theory. In this process, vital was the work of Charles Darwin. Although his book on **THE ORIGIN OF SPECIES** was published in 1859, before the term was coined it contained many seeds that could grow to dominate modern ecology.

Later, biological ecologists and population scientists used similar concepts—such as ecosystem, environmental niche (the space occupied by an organism in which it can survive and reproduce), feedback loop, stability, and growth—to address issues of population growth and environmental destruction; this line of study became particularly prominent in the 1960s and 1970s, as in the 1973 work of Paul Erhlich, Anne Erhlich, and John Holdren. Also in the 1970s, **Urie Bronfenbrenner** (1979) developed an ecological model of human development to understand the reciprocal relationships between individuals and the multiple environments in which they live. **Gerald Marten (2001)** uses human ecology and complex systems theory as a framework to examine economic systems and other social institutions and their impact on the natural environment. In particular, he discusses human ecology as a tool for resolving issues of **sustainable development and environmental problems** by understanding the complex interrelationship between human social systems and the ecosystem.

Anthropologists used ecological concepts to study the history and culture of human groups and societies to explain their success, failure, or adaptation. The concepts of equilibrium, movement of resources, sustainable development (meeting present needs without compromising the ability of future generations to meet their own needs), and the adaptation of organizational systems have been applied to the study of families, communities, race relations, schools, workplaces, government agencies, and other social institutions. It was in the study of urban environments that the concepts of human ecology achieved prominence.

Sociologists working in this area—Park and Burgess (1920s), Frazier and Sutherland (1930s), and Janowitz (1950s)—addressed key issues such as the impact of human settlement on **land-use patterns (for example, traffic flow patterns, water and flood management**), the intertwined history of industrial development and **urban decay**, race relations, and **white flight** to the **suburbs**—in other words, the components of urban and regional planning. The underlying issue understands how demands for space and resources influence the development of community and business organizations across rural and urban landscapes, and how the environment influences the structures and strictures imposed by human social systems.

Human ecology was recognized as a unique field of **geography**. Geographers went so far as declaring "**geography as the science of human ecology**." Early twentieth-century geographers sought to make clear the relationships existing between natural environments and the distribution and activities of people. However, this approach unfortunately became linked with environmental determinism which suggested that **our surroundings shape everything from skin color to behavior**. These concepts led to rather simplistic, and even racist, notions about how environments shaped cultures, and environmental determinism was discredited in the 1920s. Also during the 1920s, urban sociologists adapted ecological concepts to explain settlement patterns and human interactions in cities. **Called the Chicago School**, these sociologists adapted observation methods from anthropology to describe urban life and culture.

The sociological approach to human ecology was epitomized by the **Chicago School**—an approach to research pioneered most famously by the Department of Sociology at the University of Chicago beginning in the 1920s and 1930s—**which focused on using theory and field research methods to understand human behavior and organization in the urban environment**. Beginning in the early 1920s, scholars such as Ernest Burgess and Robert E. Park employed ecological concepts to explain the development of cities and communities. For example, the **POET model**—population, organization, environment, and technology—was developed to address the complex relationships between humans, their social organizations, and their environments. One of the architects of sociological human ecology was **Amos Hawley (b. 1910),** a population specialist and professor at the University of North Carolina, Chapel Hill, who argued that human ecology was "**the basic social science**" (1944, p. 405). Hawley is known for his work on the conceptual and theoretical foundations of sociological human ecology and the associations among population, the social-political-economic environment, and change in developing nations.

Meanwhile, the advances in geography were overshadowed by the environmental determinism criticisms. As a result, human ecology **faded** to the margins of geography and became a historical footnote in sociology for several decades. Increasingly, the social sciences became **disconnected from the physical sciences and, by extension**, from the material world. The focus of the social sciences shifted from ecological models to the embrace of economic, political, and demographic approaches where the role of natural forces was more subtle. In order to bolster the validity of their science, some researchers emphasized quantitative analysis that favored data about people over the observation of the human condition. Meanwhile, ecologists, especially those in North America, concentrated on the study of natural, non-human environments. Some one-third of the land in the United States is in public ownership, enabling wildlife and vegetation research on vast expanses with little human interruption. There are many ironies in this disconnection. For example, the Greek root for both ecology and economics is the same: oikos. Both disciplines involve the study of the household. Ecology is the study of the environmental house, including all its inhabitants, in which we live and in which we place our human-made structures and domesticated plants and animals. Economics is the study of the household of money. As we can track the flow of money, we can also illuminate other movements in the places where we live. But beyond their common Greek root, economics and ecology diverged with few clear connections persisting. Beginning in the 1960s, the general public became alarmed by **population growth and the consequences of pollution on water, air, and land quality**. **Biologists and ecologists** used human ecology to emphasize how people are subject to the same environmental limitations as other animals. Also during this time, anthropologists used the term to help explain the impact of **environment on culture**. Ecologically oriented anthropologists adapted concepts like population regulation and energy flow to explain community organization. In general, early use of ecological concepts in human ecology depended on traditional views of nature, such as the tendency of systems to evolve toward a steady state. This past suggests the ongoing utility of human ecology. By understanding the interactions and interrelationships between people and their environments, human ecology can help to:

- Consider and plan for the long-term consequences of human actions;
- Avoid **disastrous surprises** resulting from environmental phenomena such as floods, earthquakes, tornadoes, wildfires, and tsunamis;
- Generate ideas for dealing with environmental challenges and opportunities; and
- Create a livable and sustainable relationship with the environment. However, to realize this utility, it is necessary to understand changes in ecological thinking generally and how.

Ideas Contributing to a New Human Ecology

Human Ecology lends itself to reinvention, to reinterpretation. Relationships link things, and how we view connections among elements changes. As early as the 1950s, anthropologists called for a '**new ecology**'. The ideas leading to the more recent, expanding view of ecology have come from many sources and a variety of disciplines, including anthropology. **The catalysts for change include advances in technologies,** the study of **urban morphology** and **landscape ecology**, a broader understanding of chaos theory, and increased interest in issues of sustainability. The emergence of urban ecology exemplifies a beginning in the synthesis of these sometimes divergent catalysts. Urban ecology focuses on organism–environment interactions within cities and other human settlements. By concentrating on urban areas, the interests of the new ecological perspective are woven closer together. Human ecology has a history of focusing attention on humans' impact on the biotic world. **Paul Sears** was an early proponent of applying human ecology, addressing topics aimed at the population explosion of humanity, global resource limits, pollution, and published a comprehensive account on human ecology as a discipline in 1954. He saw the vast "explosion" of problems humans were creating for the environment and reminded us that "what is important is the work to be done rather than the label.

Fresh ways to observe nature, primarily as a result of computer and remote-sensing technologies, have altered our understanding of functions, structures, and patterns. These new (and evolving) technologies are yielding a deeper perspective, because many events can be considered simultaneously in a connected network.

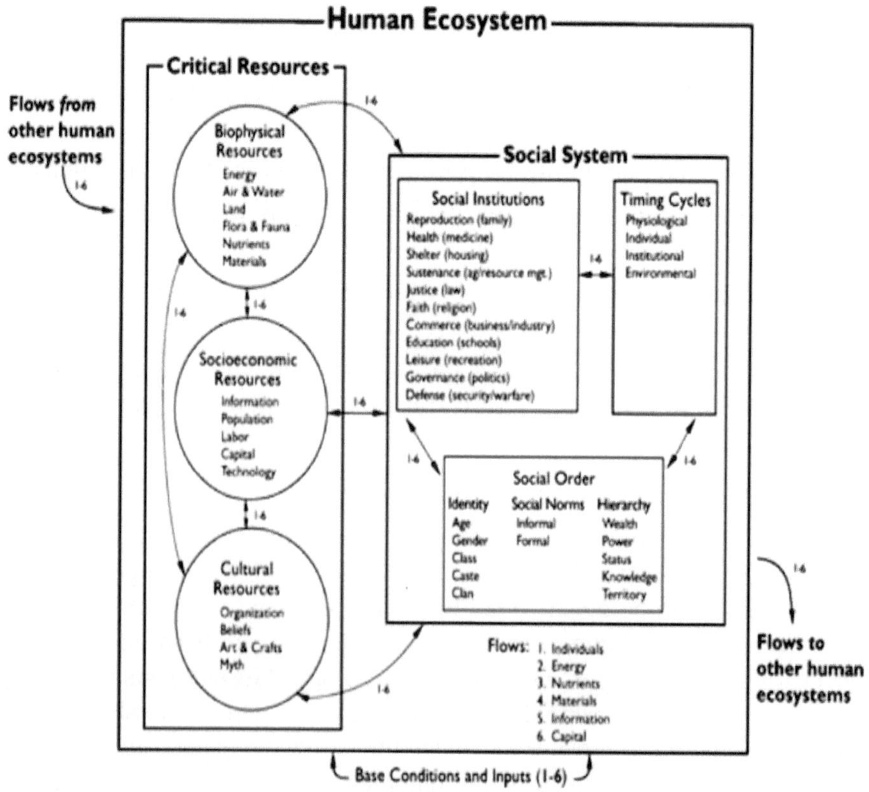

Facuilty of human ecology

a) Geographical Information Systems (GIS)

A computer technology especially valuable for revealing complex, ecological relationships is geographical information systems, known by its abbreviation GIS. These computer software programs allow analysis to study overlapping spatial data and map the results. For example, the home range of a tiger beetle species can be mapped then compared with a similar map for a species of brown bear. In turn, both can be overlaid on the migration routes of Canada geese and the extent of a coniferous forest and so on. GIS emerged concurrently with new ways to see and to record the surface of the planet, such as remote-sensing technologies. Whereas GIS programs map information, remote sensing creates imagery of phenomena on the Earth's surface. Continents and water bodies were clearly visible beneath swirls of clouds, but borders had disappeared. The images can be enhanced with computers to reveal specific phenomena, such as land cover, land use, and fault lines. Climate patterns can be tracked and future weather events forecasted. Remote sensors can also be linked to on-the-ground monitoring stations. Such connections allow phenomena to be observed through time. For example, a drainage basin can have several stream-monitoring gauges, which may be linked to a central data collection center. In turn, satellites may be able to collect rainfall and snowpack information daily that can be combined with the field data to predict future water supplies.

b) Nested Networks

Living systems are organized hierarchically and communicate through feedback networks. The elements within a system may vary greatly in organization. According to urban morphologists, urban form can be understood at different levels of resolution. Commonly, **four** are recognized, **corresponding to the building/lot, the street/block, the city, and the region**. A building may be tall or short, with a pitched roof or a flat one, brick or wood or adobe. A lot, a street, or a block may be narrow or wide, straight or curved. A city may be densely settled or spread out. A region can be defined by a river or a mountain or a coastline or all three and by other factors. Hierarchies help us understand how people are connected with one another – the basic idea of community. To understand human ecologies, the most relevant levels of organization include habitat, community, landscape, region, nation and state, and Earth or ecosphere. These levels present different, yet interconnected, scales of analysis. Each level possesses a history and a literature of analysis and debate. The habitat includes the building and lot. The community is comprised of buildings, lots, streets, and blocks. Landscapes can be urban, suburban, rural, and wild. Regions are hodgepodges of landscapes, while the distinctions between regions, and often those between states and nations, and are even more blurred. But there is less ambiguity about the ends of the Earth. Each level of human organization (nation, state, region, county, and city) is an element in a larger system, but is also comprised of smaller geographic units like neighborhoods and communities, which are, in turn, collections of single households. Home and work places form the habitats for people and are further divided into cells we call rooms. Hierarchy may be seen as a framework, a system of nested networks. A critical feature of these nested networks is an asymmetric interaction in between levels. The larger, slower levels maintain constraints within which faster levels operate. There are, however, circumstances when slower and larger levels in ecosystems become briefly vulnerable to dramatic transformation because of small events and fast processes. Large, slow levels tend to keep things in place. Small, fast levels initiate changes when the larger levels are not functioning effectively. New ecology, a deeper understanding of interactions at various scales, holds the prospect for better, although more complex, approaches to sustainable resource management, nature conservation, and environmental protection as well as the arts of environmental design and planning.

The new human ecology explores not only the influence of humans on their environment but also the influence of the environment on human behaviour and their adaptive strategies as they come to understand those influences better. It is a way of thinking about the world, and a context in which we should define our questions and ways to answer those questions. Human ecology is the sub-discipline of ecology that focuses on human. Human ecology studies the relationship between humans and the natural, social and built environment.

The main tasks of human ecology are

i) The investigation of human health condition.

ii) The forecast of the health condition of the future generations

iii) The research of dynamics of human health condition in the process of historical and social – economical development

iv) The investigation of the processes of human health protection

v) The analysis of global and regional problems of human ecology

vi) The research of influence of environmental factors to human health.

vii) The composition of medical geographical maps that show the territorial differentiation of human diseases.

viii) The addition of medical geographical maps and environmental pollution maps & the determination of correlative dependence between human diseases and environmental pollution.

ix) The determination of value of boundary technogenic load limit to human organism.

The mankind expands areas of **dwelling by building megalopolises with skyscrapers of height exceeding 800 meters and going deep into Earth subsoil by means of mines and wells for production of energy and raw material resources.**

Constantly growing requirements of human society determine emergence of more and more effective technical means and technologies for development of raw fields. However, an increase in the rates of an expenditure and production of resources, which is inconsistent with the possibilities of self-restoration of Earth's geospheres, leads to violation of natural balance. By the beginning of the 21st century, influence of a technogenesis has reached the menacing scales, and a danger of death of all life on Earth became obvious to a considerable part of the population.

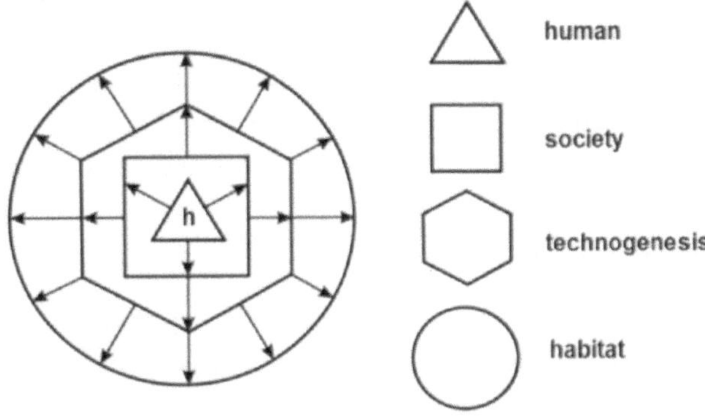

SCOPE OF HUMAN ECOLOGY

Human ecology has been defined as a type of analysis applied to the relations in human beings that was traditionally applied to plants and animals in ecology. Toward this aim, human ecologists (which can include sociologists) integrate diverse perspectives from a broad spectrum of disciplines covering wider points of view. Scopes of topics in human ecology are:

- The role of social, cultural, and psychological factors in the maintenance or disruption of ecosystems;
- Effects of population density on health, social organization, or environmental quality;
- New adaptive problems in urban environments;
- Interrelations of technological and environmental change
- The development of unifying principles in the study of biological and cultural adaptation
- The genesis of mal-adaptations in human biological and cultural evolution;
- Genetic, physiological, and social adaptation to the environment and to environmental change;
- The relation of food quality and quantity to physical and intellectual performance and to
- demographic change;
- The application of computers, remote sensing devices, and other new tools and techniques

While theoretical discussions continue, research published in Human Ecology Review suggests that recent discourse has shifted toward applying principles of human ecology. Some of these applications focus instead on addressing problems that cross disciplinary boundaries or transcend those boundaries altogether. Human ecology is neither anti-discipline nor anti-theory, rather it is the ongoing attempt to formulate, synthesize, and apply theory to bridge the widening schism (disagreement) between man and nature. This new human ecology emphasizes complexity over reductionism, focuses on changes over stable states, and expands ecological concepts beyond plants and animals to include people.

Applications of Human Ecology

1. **Application to epidemiology and public health**: The application of ecological concepts to epidemiology has similar roots to those of other

disciplinary applications, with Carl Linnaeus having played a seminal role. However, the term appears to have come into common use in the medical and public health literature in the mid-twentieth century. This was strengthened in 1971 by the publication of Epidemiology as Medical Ecology, and again in 1987 by the publication of a textbook on Public Health and Human Ecology. An "ecosystem health" perspective has emerged as a thematic movement, integrating research and practice from such fields as environmental management, public health, biodiversity, and economic development. Drawing in turn from the application of concepts such as the social-ecological model of health, human ecology has converged with the mainstream of global public health literature.

2. **Connection to home economics**: In addition to its links to other disciplines, human ecology has a strong historical linkage to the field of home economics through the work of Ellen Swallow Richards, among others. However, as early as the 1960s, a number of universities began to rename home economics departments, schools, and colleges as human ecology programs. In part, this name change was a response to perceived difficulties with the term home economics in a modernizing society, and reflects recognition of human ecology as one of the initial choices for the discipline which was to become home economics.
3. **Ecosystem Services**: The ecosystems of planet Earth are coupled to human environments. Ecosystems regulate the global geophysical cycles of energy, climate, soil nutrients, and water that in turn support and grow natural capital (including the environmental, physiological, cognitive, cultural, and spiritual dimensions of life). Ultimately, every manufactured product in human environments comes from natural systems. Ecosystems are considered common-pool resources because ecosystems do not exclude beneficiaries and they can be depleted or degraded. For example, green space within communities provides sustainable health services that reduces mortality and regulates the spread of vector borne disease. Research shows that people who are more engaged with regular access to natural areas have lower rates of diabetes, heart disease and psychological disorders. These ecological health services are regularly depleted through urban development projects that do not factor in the common-pool value of ecosystems. The ecological commons delivers a diverse supply of community services that sustains the well-being of human society. The Millennium Ecosystem Assessment, an international UN initiative involving more than 1,360 experts worldwide, identifies four main ecosystem service

types having 30 sub- categories stemming from natural capital. The ecological commons includes **provisioning** (e.g., **food, raw materials, medicine, water supplies**), **regulating** (e.g., climate, water, soil retention, flood retention), **cultural** (e.g., science and education, artistic, spiritual), and **supporting** (e.g., soil formation,nutrient cycling, water cycling) services.

Interdisciplinary Approaches

Human ecology expands functionalism from ecology to the human mind. People's perception of a complex world is a function of their ability to be able to comprehend beyond the immediate, both in time and in space. This concept manifested in the popular slogan promoting sustainability: "think global, act local." Moreover, people's conception of community stems from not only their physical location but their mental and emotional connections and varies from "community as place, community as way of life, or community of collective action." In these early years, human ecology was still deeply enmeshed in its respective disciplines: geography, sociology, anthropology, psychology, and economics. Scholars through the 1970s until present have called for a greater integration between all of the scattered disciplines that has each established formal ecological research.

1. **In Art:** While some of the early writers considered how art fit into a human ecology, it was Sears who posed the idea that in the long run human ecology will in fact look more like art. Bill Carpenter (1986) calls human ecology the "possibility of an aesthetic science," renewing dialogue about how art fits into a human ecological perspective. According to Carpenter, human ecology as an aesthetic science counters the disciplinary fragmentation of knowledge by examining human consciousness.
2. **In Education**: While the reputation of human ecology in institutions of higher learning is growing, there is no human ecology at the primary or secondary education levels. Educational theorist Sir Kenneth Robinson has called for diversification of education to promote creativity in academic and non-academic (i.e.- educate their "whole being") activities to implement a "new conception of human ecology"
3. **In anthropological perspective**: While cultural beliefs come into focus in the influential modern sciences like economics, human ecologists examine the modern concepts of economic growth and technological development from an anthropological perspective. By comparing those concepts with other scientifically insights about environmental degradation, climate change and global inequality, human ecology

challenges the ideas that perpetuate an unsustainable and unequal global society. Studies in Human Ecology give you a broad and theoretically deep understanding of the interactions between man and nature in different times and in different parts of the world. Of central importance is to understand how the human relationships with the environment are influenced by their history and their place in the world system.

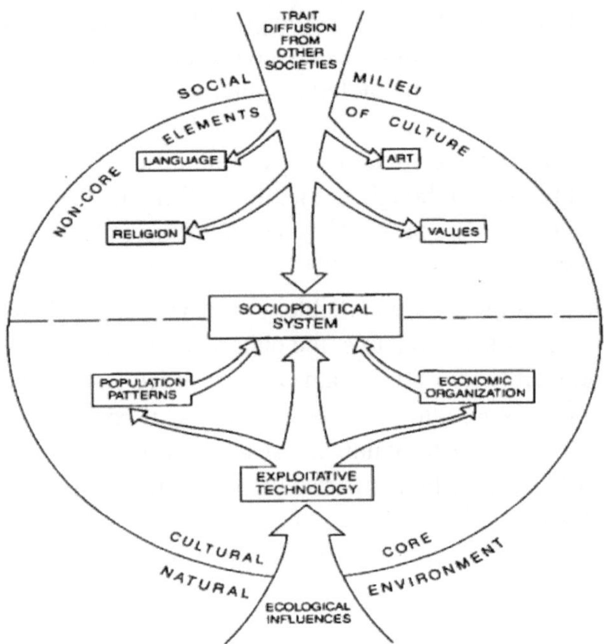

1. **Interpretations of Ecology:** Interpretations of the concept and substance of ecology differ according to the unique advantage points provided by independent scholarly traditions.

Conceptual approaches to human ecology

The concept of Cultural Ecology

Theories of human interaction provides a way of making sense of events that have happened in the past, and then allow us to make predictions about what may happen in the future. ***Human ecology theory*** is a way of looking at the interactions of humans with their environments and considering this relationship as a system. (Fig. Models of cultural ecology) In this theoretical framework, biological, social, and physical aspects of the organism are considered within the context of their environments. These environments may be the natural world, reality as constructed by humans, and/or the social and cultural milieu in which the organism exists.

The Biological Interpretation

Ecology was originally given a biological interpretation in 1866 when Ernst Haeckel defined the science in accord with the work of Darwin, who had attempted to explain evolution in terms of the adaptations of organisms to the biotic and abiotic features of their environments (Fig: Ecosystem based model of human ecology). A generation later another foundation concept appeared: since evolutionary studies indicated there was ecology of living organisms before humans evolved, and humans themselves are living organisms, there must therefore be a human ecology. On the biological interpretation, humans are part and parcel of the natural world, and are subject to the same biophysical dynamics that govern the organization of other living communities. For most of the twentieth century, however, although gradually shifting in the last couple of decades, biologists took only a passing interest in the scientific study of the ecology of humans per se.

However, our internal biology is quite unremarkable. For example, we share about 99% of our DNA with our closest relative, the chimpanzee. Our bodies metabolize energy at a predictable rate for a mammal of oursize (Ulijaszek 1995) and, like all other organisms, our diets must maintain a predictable balance of chemical elements in order to maintain our internal biochemistry (Williams and Frausto da Silva 2006). In comparison to other mammals, our life history is quite predictable for a primate species of our body size , with some unique features, such as a particularly long lifespan, large encephalization, a significant post-reproductive lifespan in human females, and the requirement of large amounts of parental investment in our highly altricial, slow-growing offspring (Charnov and Berrigan 1993; Hawkes *et al.* 1998;Kaplan *et al.* 2000; Walker *et al.* 2006).

One of the most striking biological features of the human species is the large size of our brain, a powerful biophysical computer that allows us to absorb, process, interpret, and store large quantities of information about the external world (Eccles 1991; Schoenemann2006, 2009). These abilities have led to a sophisticated understanding of basic chemical, physical, and bio-logical principles, and have allowed us to use these principles to engineer solutions to many of the energetic and informational constraints faced by other bio-logical species, beyond the genetic pathways shared with other organisms (Williams and Frausto da Silva 2006). The immense human cognitive capacity resulted in the evolution of a key secondary pathway of heritable information some time in our evolutionary past, which has ever since been fundamental to human biocultural evolution.

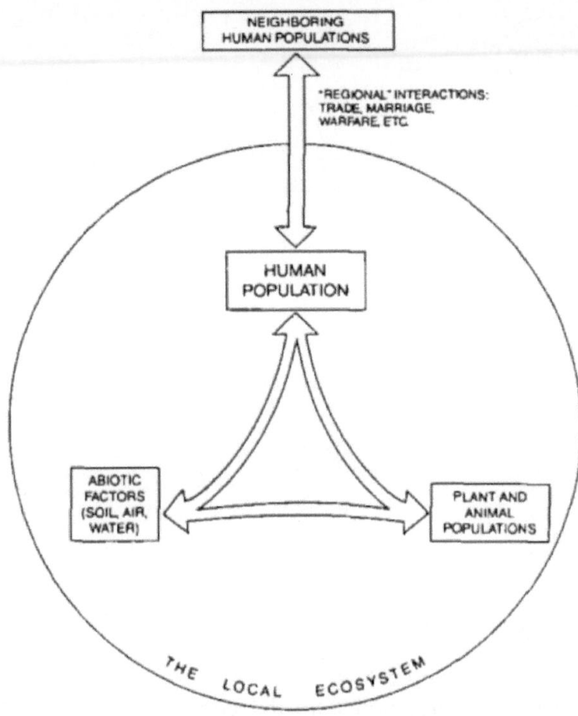

The actor-based model of human ecology

In the face of severe empirical problems in defining the social unit of ecological adaptation, it has been suggested that adaptation occurs at the level of individuals rather than of cultures or populations. This actor-based model of human ecology, as Orlove (1980) has labeled it, has become the major new wave in human ecology. The model reflects both anthropologists' general concern with individual decision-making processes and evolutionary biologists' current preoccupation with showing that natural selection operates exclusively at the level of the individual organism. From this perspective, any higher levels of organization, whether communities, ecosystems, or human social systems, exist only as the fortuitous outcome of interactions among many individual organisms.

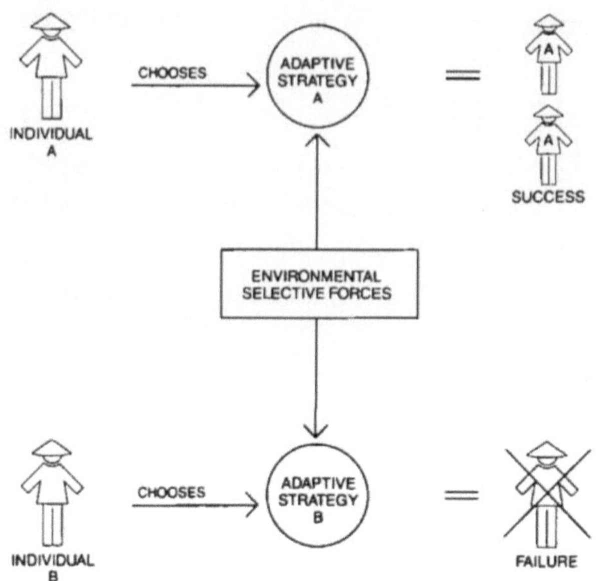

The systems model of human ecology

A major scientific development in recent years has been the formulation of "general systems theory," which is concerned with the general properties of the structures and functions of systems as such, rather than with their specific contents. According to this theoretical approach, atoms, cells, organisms, ecosystems, societies, and even the universe as a whole all share the common propertics of being self-organizing systems and can therefore be studied in terms of a common theoretical perspective.

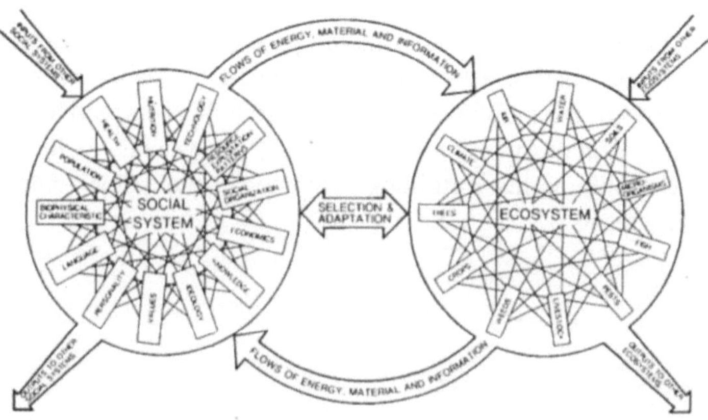

Figure 6. Social system–ecosystem interactions.

The Metaphorical Interpretation

Social scientists, by contrast, were quick to seize on the idea of a human ecology with independent investigations begun in the 1920s, in India, by Radhakamal Mukerjee, who followed the biological interpretation, and in the USA by Robert Park, whose students did not. Aware of theories of community succession that their biological colleagues were then developing for natural communities, in which society was used as a metaphor to understand natural organization, some early US sociologists came to use nature as a metaphor to understand social organization. Setting aside the rhetorical uses of 'ecology' as a synonym for 'contextual' or 'environmental,' two variants of metaphorical interpretation can now be found in social science. **The first appliesbio-ecological concepts** to the study of human communities per se, reasoning by analogy. Here, institutional or organizational settings are taken as contexts or environments for humans within their groups, populations, or societies, which are supposed to be organized parallel to natural communities. This variant generally overlooks the fact that natural communities contain multiple species and humans are a single species, preferring to finesse this by elaborating biological species and evolutionary analogies for societal divisions of labour. **The second variant** treats ecological knowledge as a social construct (ideas constructed by the researchers). Left unaccounted for in the metaphorical interpretation are biological connections between human societies and other natural systems. The metaphorical interpretation is a blueprint for the study of the internal workings of society through the adaptations of its populations or groups by means of their social organizations and shared cultural knowledge.

The Ideological Interpretation

With the reawakening of the modern environmental movement world wide and its growth in the last decades of the twentieth century, an ideological interpretation of ecology has appeared. In this, the biological and metaphorical interpretations of ecology are freely interwoven and supplemented with value judgments gauged against standards supposedly congruent with ecological science. Some versions of the ideological interpretation use nature as an ideal to evaluate and reform **social and political organization, economic practice, and personal conduct,** while others try to frame **environmental ethics** so as to integrate ecological science with politics, economics, philosophy, education and various religions and cultures. The ideological interpretation is neither monolithicnor uniform, but its advocates generally fall to the left of ideological spectra. The common thread is that **ecology is more than science or metaphor**: It is a worldview with ethical implications for personal, cultural,and institutional change.

Theory Frames for Human Ecology

The multifarious research traditions of human ecology may be understood in terms of the different theory frames they have employed, which reflect their respective interpretations of ecology. Different theory frames address different **problematics**.

1. The Evolutionary Theory Frame

The evolutionary theory frame assumes that ecological forces act on genetic and cultural variation among individuals and populations so as to effect a selective survival and reproduction. Using this assumption, neo-Darwinian principles of natural selection are applied to understand how individuals **adapt** to their natural environments, how different social species reflect common principles of social organization in response to ecological contingencies, and how populations and cultures are affected by pathogenic disease. Individual behaviour and social organization are classic problematic in social science. The application of evolutionary theory attempts to provide a fuller explanation of these. One strand of theory develops a human evolutionary ecology patterned on the behavioral and evolutionary ecology of the biological sciences. In this, an adaptive design is assumed to be the product of the **elemental behaviour of individuals**, such as foraging, territorial defense, time allocation and mate selection, all of which are presumed to satisfy economic cost-benefit principles (Smith and Winterhalder 1992). It has yet to be demonstrated whether neo-Darwinian evolutionary-economic models have application beyond the case of hunter-gatherers. Because organic traits (such as brain size, bipedalism, and vocal chords) underlie cultural capabilities, and culture is understood as a human adaptation, some evolutionary models assume the coevolution of biological and cultural traits in ecological context (Boyd and Richerson 1985, Durham 1991). Another strand of theory focuses on problematics of social organization among different species, with a view to discovering common principles but without presuming that the properties of social organization can be reduced to the **summed behaviour of individuals**. In addition to comparisons with social primates, a research tradition of long standing, Machalek's (1999) transpecific sociology enables the comparison with humans of the demographic constraints, division of labour, and homeostatic mechanisms that appear in colonies of social insects, and of common strategies of environmental exploitation across social species.

Problematics of epidemiology (Challenges of epidemiology), ancient, and contemporary, fall within an evolutionary theory frame in as much as most scholars consider pathogens to be the principle agents of natural selection operating on humans since sedentary lifestyles were adopted 10,000-12,000 years ago. In modern times, medical authorities are again sounding alarms about the development and spread of epidemic pathogens, old and new, because normal genetic mutations eventually produce drug resistant variants of old as well as new, and now dangerous, variants of formerly benign organisms. Since microbic pathogens are part of the living environment of all humans, human culture and society have to adapt to ever-changing conditions of pathogenic disease just as pathogens adapt to humans—an inherently ecological condition. The incidence of pathogenic diseases is known to be affected by agricultural practices, dietary customs, migration patterns, population densities, climate alteration, cultural hygiene practices, lack of potable water, sexual mores, modern transportation that transcends normal environmental barriers, and technological methods of intervention (Ewald 1994, Kipple 1997).

Intervention methods typically disrupt population equilibria that might normally exist between pathogens and humans, thus preventing natural environmental controls from exerting themselves. Instead of stabilizing natural environments, technology implemented through the social organization of medicine and public health rather attempts to control pathogens so as to stabilize their relationship to humans. This method in itself, by intervening to change natural biota, provides a stimulus for microbic evolutionary adaptation. The general principle is that changed environments bring changed frequencies of microbic pathogens, human intentions notwithstanding. Some scholars (Mascie-Taylor 1993) have tended to conceive human ecology entirely in these epidemiological terms; however, this has not been predominant.

2. The Environmental Theory Frame

The environmental theory frame dominates modern research traditions of human ecology. It too presumes a biological interpretation of ecology, but in some cases substantially supplemented by considerations of social science targeted to problematics of subsistence and sustainability. Anthropologists and geographers have a long tradition of investigating how human behaviour, culture, traditions, settlements, built environments, spatial location, territory, and humans' sense of place are affected by efforts to procure subsistence from natural—typically regional—environments. Research programs investigating **cultural ecology, ecological anthropology, environmental geography, and environmental history** consistently demonstrate how the development of culture and society vary with settlement patterns, technology and means of material resource acquisition and how natural environments are modified by socio-cultural variations in subsistence practices. A ***cultural materialism***, which viewed culture and its symbolism as epiphenomenal, was implicit or explicit in many of these classic approaches, as was a dualistic construal of

nature and culture. Modern approaches, especially political **ecology**, are different (Bier sack 1999, Scoones 1999). They focus on the interaction of spatial and temporal (sequential) environmental dynamics with people situated in political, social, institutional, and cultural context, and emphasize the effects of human agency within the push-pull parameters of social, economic and political inequalities, on the one hand, and local or regional environmental transformations, on the other.

Political ecology conceives human-environment interactions as mutual, thus linking human ecology to political economy and crossing from problematics of subsistence to those of sustainability (Feminist Political Ecology).No problematic has so dominated modern environmental discourse as the question of limits to growth, specifically, whether industrialism is sustainable into the indefinite future as it is adopted round the world. Biologists began the debate in the 1960s by importing from population ecology concepts of carrying capacity limits, and they averred that cascading world population growth augured for **environmental catastrophe**. Some social scientists amplified this **neo-Malthusian** argument, noting that industrialism is not a population but a way of life, one which produces too much consumption not just too many consumers. Attentive to **anthropogenic environmental degradation** (aquifer depletion, biodiversity loss, deforestation, ozone depletion, global warming, etc.), the new social science tradition of examining environmental interfaces with human production and reproduction is now secure with the appearance of numerous textbooks and treatises on environmental sociology, environmental psychology, and ecological economics. These specialties tend to conceive human-environment interactions within the existing theory traditions of their established disciplines. Recently, ominous neo-Malthusian analyses have been challenged by more optimistic theories of ecological modernization (Spaargaren and Mol 1992).

Theory aside, problematics of sustainability are routinely encountered as matters of practice in the concerted effort to promote sustainable development in economically less secure nations of the world. Specifically promoted by the United Nations, sustainable development is taken to mean eliminating world poverty while not compromising local environmental stabilities (Brundtl and Commission 1987). In effect sustainable development programs have promoted further industrialization—thus rekindling the question of the long term sustainability of human-environment interactions as local, regional, and global environmental degradation compounds, interactively cascading from industrialism's increased energy capture and waste disposal. The concern over environmental degradation directs the problematic of sustainability, which currently is targeted more towards local than toward global applications.

3. Theory of Environmental Determinism

Around the turn of the century, geographers, notably Friedrich Ratzel in Germany and his American disciple, Ellen C. Semple, came with the view that humans were completely the product of their environment, a theory that came to be called environmental determinism. Followers of this school, which dominated geographical thought well into the 1920s, asserted that all aspects of human culture and behavior were caused directly by environmental influences (Figure 1).

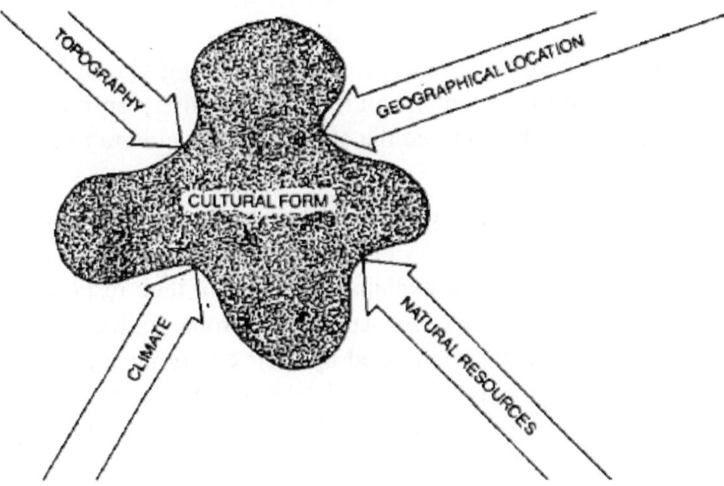

Figure 1: The model of Environmental Determinism (Source: Rambo, 1983)

For example, the British were a nation of seafarers because they were an island-dwelling race surrounded by seas; the Arabs were monotheistic Muslims because living in the vast empty desert turned their minds toward a single God; the Eskimos were primitive nomads because the harsh conditions of their arctic habitat forbade their development into a complex civilization. There is simply too much variation in human behaviour in seemingly similar geographical settings for it to be environmentally determined.

Environmental Determinism definition: A deterministic approach assigns one factor as the dominant influence in explanations. Environmental determinism is based on the assumption that cultural and natural areas are coterminous, because culture represents an adaptation to the particular environment (Steward, 1955). Therefore, environmental factors determine human social and cultural behaviour (Milton, 1997).

4. Theory of Environmental Possibilism

In place of the discredited determinism, a new theory, called environmental possibilism, was proposed. Its proponents asserted that while the environment did not directly cause specific cultural developments, the presence or absence of specific environmental factors placed limits on such developments by either permitting or forbidding their occurrence (Figure 2). Thus, island people could be seafarers, but residents of Inner Mongolia could not be; inhabitants of temperate regions might practice agriculture, but those living in arctic latitudes could not. The value of the possibilist approach was perhaps best demonstrated by the American anthropologist Kroeber, who showed that the Indians of northwestern North America could not adopt maize agriculture from their southern neighbors because the frost-free growing season in their region was shorter than the four months required for the maize plants to reach maturity. Their environment thus limited the ability of their culture to evolve in an agricultural direction.

A possibilist stance was also taken by the British historian Arnold Toynbee in his multi volumed A Study of History (1947), in which he argued that the development of civilizations could be explained in terms of their responses to environmental challenges. Cultures located in the benign tropics failed to evolve because they were not sufficiently challenged by their environment; those in extremely harsh habitats such as the Eskimos in the arctic remained forever primitive because simply coping with the demands of their environment sapped all of their creative energies. Only those cultures in environments offering sufficient but not excessive challenges had the possibility of progressing to higher stages of civilization.

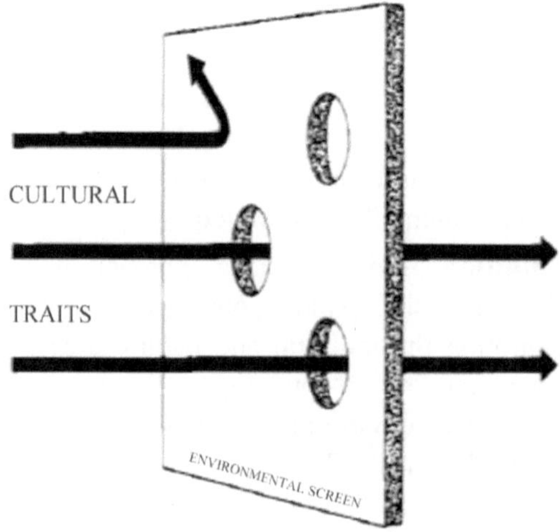

Possibilism suffers from one overriding defect as a scientific theory; it lacks any general predictive or explanatory power since it is able to explain only why certain developments could not occur in certain environments. It is totally unable to predict whether or not they would occur under favourable circumstances. For example, the failure of Eskimos to grow corn is explainable, but possibilism cannot explain why the English were great seafarers while the Tasmanians were not. Clearly, the difference in the latter case was due to existence of very different cultural traditions and bodies of technological knowledge rather than reflecting environmental influences. In short, as the British anthropologist Daryll Forde concluded in his book, Habitat, Economy and Society (1934), which was perhaps the last major scientific exploration of possibilism, "between the physical environment and human activity there is always a middle term, a collection of specific objectives and values, a body of knowledge and belief: in other words, a cultural pattern." (The model of Environmental Possibilism (Source: Rambo, 1983). With this realization, social scientists tended to turn from studying human interactions with the environment, preferring instead to focus on the seemingly more profitable study of the internal structure and functioning of cultural and social systems.

5. The Analogic-Symbolic Theory Frame

This theory frame argue that analogical reasoning, particularly Gentner's (1983, 2010) structure-mapping theory, provides an integrative theoretical framework through which we can better understand the development of symbol use (symbolic representations, scale models, photographs, and maps are used for what intention, comparison, and similarity). Analogical reasoning can contribute both to the understanding of others' intentions and the establishment of correspondences between symbols and their referents, two crucial components of symbolic understanding. The analogic-symbolic theory frame works with a metaphorical interpretation of ecology, with different variants selectively emphasizing analogies or symbolic constructs as central to theory development. Generally speaking, the theory frame takes biological ecology just seriously enough to borrow some of its content (sometimes merely its words) so that nature can serve as a metaphor to understand society, thus advancing social science. In the classic sociological tradition, 'environment' usually had no biological referent, and neither does 'community' or 'ecosystem,' which is favorite analogs in modern sociological human ecology. As noted earlier, US sociology provided the first sustained research tradition of human ecology, something of an urban geography, concerned as it was with spatial relations in cities beginning in mid-twentieth century.

In this, territorially circumscribed human populations are characterized by technologically based organizational adaptations in the sustenance producing activities of populations within their social, political, and economic environments. These orienting concepts have not changed in more than a generation (Micklin and Posten 1998). They reflect a social and demographic, but not a biological, materialism in the conception of purely sociological problematics. Direct descendants, such as organizational ecology, and distant cousins, such as family ecology, likewise depend on theoretical analogies to biological phenomena in order to relate social entities to social environments.

6. The Interactive Theory Frame

The interactive theory frame for human ecology is the newest, most diverse and most numerous in its scholarly problematics. Its growing traditions may only be listed here. They include assorted environmental ethics, such as deep ecology, the land ethic, and animal rights; assorted spiritual ecologies, including native US land wisdom, Christian-based ecological process **theology**, and **strains of Buddhist**, Taoist, and Jainist thought; some applications of complexity theory for evolving interactive systems; Gaian ecology;politically grounded approaches, such as social ecology and green Marxism; eco-feminism; and more.

All of these assume some variant of the ideological interpretation of ecology, reflecting a wish to integrate ecological knowledge with selected priorities in human affairs. Some are based in science, while others are based in philosophy, politics, religion, or tradition, but almost none maintains a hard and fast distinction between objective and subjective knowledge. **Almost all take a holistic rather than reductionists approach to their problematics,** in which the traditional boundaries of scientific disciplines, and even the classic boundaries between science and the humanities, are regarded as anachronistic. The overriding theme construes human ecology as knowledge of the complex interactive connections that integrate human activities with environmental contexts. The overall result is an eclectic(diverse) combination of biological, environmental, moral, and even aesthetic, approaches to human ecology.

1.2 Ecology of developmental processes. Bio-ecological model of Bronfenbrernner

Ecological Theory of Development

There have been many different approaches to explaining development, but one especially has tried to deal with the importance of the environment without ignoring the uniqueness of the individual. Urie Bronfenbrenner, and more recently along with Pamela Morris, has developed what is called an experimental ecology of human development or an ecological theory of human development. Within this model he stresses the importance of the developing person in his or her surrounding environment. He defines the phrase ecology of human development as the study of "**the progressive, mutual accommodation, throughout the life span, between a growing human organism and the changing immediate environments in which it lives**."

His basic argument is that traditional studies in human development are very rigorous and tightly controlled. While this can be a benefit, the studies are also very limited in scope because many of these experiments take place in settings that are unfamiliar to the participants and artificial in their construction. In other words, they don't very accurately represent what the real world is about; in Bronfenbrenner's earlier and own words, developmental research has been "the study of the strange behavior of children in strange situations for the briefest possible period of time." Because he is so concerned with the qualities and characteristics of the environment, part of his ecological model defines a series of structures that are "nested" within one another. In his earliest works, he described four such structures. These structures nest or "fit" within each other, beginning with the microsystem (as he points out much like the stacking Russian dolls that disappear within one another).

This early model of the ecology of human development helps us in two primary ways as far as understanding human development. First, it places the interaction between nature and nurture in a very clear and easily definable context of one of the four systems discussed above.

Second, it encourages us to move away from laboratory-based settings and begins to examine development in the "natural stream" of when and where it occurs. More and more scientists who study human development are emphasizing the qualitative nature of changes and using such methods that reflect that emphasis—away from the more tightly controlled laboratory study.

But, as with all good scientists, Bronfenbrenner has moved on. Along **with Morris's help**, he has taken the next step and further developed their ideas placing an emphasis on several new and complimentary ideas. The primary change in the ideas presented almost 20 years ago, and now referred to as part of a bioecological model, involves proximal processes (or interactions in the immediate environment).

Bronfenbrenner's original model has clearly expanded to further emphasize the importance of the environment and the interaction between the individual and his or her environment. It is a rich display of ideas and the importance of the role of our social world and its influence upon us.

The Life-Course and Human Development: An Ecological Perspective

- *Ecological framework*—Framework that considers the systems in which humans are embedded rather than in isolation from context
- *Enabling environments*—The systems, forces, and factors that create the conditions in which optimal child and adolescent development is possible
- *Life course perspective*—Acknowledgment that each stage of development builds on the last, focusing on the full continuum of development from preconception through adolescence and beyond
- *Biological embedding*—Early developmental contexts cause stable epigenetic modifications in the individual and these modifications endure, shaping how the individual responds to subsequent exposures over the life course
- *Developmental cascade*—Early functioning in one domain of behaviour or emotional functioning spills over to influence other domains

A life-course perspective is applied to the study of human development in ecological context. Three meanings of age (developmental, social, and historical) represent key elements of this perspective and depict lives in terms of aging, career, and historical setting. Age locates people in history (by birth year) and in the social structure. The neglect of such temporal distinctions in problem formulation has consequences in studies of status differences and psychological states, of careers and work satisfaction, of children's socioeconomic environment and the family economy, and of life change and stress. Alternative questions based on the life-course facilitate explanatory assessments of the relation between environmental and personal change.

There is a synergy between ecological and life course approaches to child health and development. Life course approaches complement ecological models in describing how individual life experiences are shaped by their immediate and extended environments, by adding the dimension of time, showing how current child and adolescent functioning depends in part on past exposures and experiences.

An ecological life course model encourages us to think about development as a process that is **cumulative and continuous**, and cautions against approaches to programming that silo risks or age groups or that view children and adolescents in narrow "disease model" terms. In ecological life course perspectives, what happens to a child at each developmental stage is shaped and influenced by what has happened in all earlier stages. Child and adolescent health and development are interdependent and consecutive, with progress in one period being shaped by influences and events in preceding periods. In addition, an integrated ecological life course approach implies that intervention efforts should be different for different populations, determined by the unique combinations of risk and protective factors to which they are currently, and have previously been, exposed.

At the individual level, much is known about the relation between risk and protective factors and health and development outcomes. However, there is a social gradient for risk and protective factors, as well as outcomes. The concepts of biological embedding and developmental cascades describe the mechanisms by which social gradients in health and illness come to exist within an individual—how they get "under the skin."One example is that low socioeconomic status is a risk factor for exposure to violence. Children exposed to violence early in life may embed (through neurological factors and endocrinology) a heightened sensitivity to danger cues in the environment. A child with anxiety may then enter school, develop further internalising symptoms, be distracted from work, and receive low grades. This hypothetical child is then more likely to remain in a low socioeconomic environment, as options for employment are curtailed by educational failure.

Bronfenbrenner's Ecological Theory Model

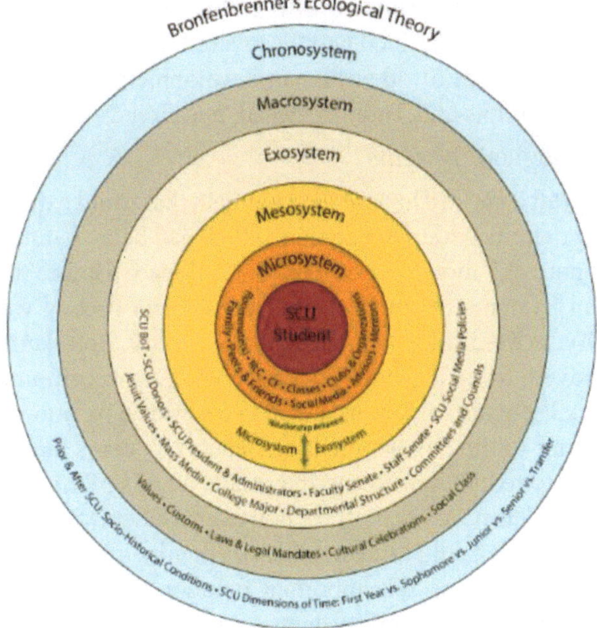

The Ecological Systems theory represents a convergence of biological, psychological, and social sciences. Through the study of the ecology of human development, social scientists seek to explain and understand the ways in which an individual interacts with the interrelated systems within that individual's environment. Bronfenbrenner (1994) sought to develop a theory of human development that would consider the influences of all of the systems that play a role in impacting the lived experiences of the individual no matter how remote the influence. Today, many of these concepts are commonplace in social work practice (micro-, meso-, and macrosystems, for instance); however, its foundation of empirical support is often less understood.

Ecological Systems theory describes human development through the prism of the "environmental interconnections and their impact on the force directly affecting psychological growth". Through the study of the ecology of human development, social scientists seek to explain and understand the ways in which an individual interacts with the interrelated systems within that individual's environment. The theory represents a convergence of biological, psychological, and social sciences. According to Bronfenbrenner, "human development is the product of interaction between the growing human organism and its environment" with the developing person seen as **malleable** within the social milieu of his or her environment. The change needs to be experiential and lasting in order for development to occur. The Ecological Systems theory provides a theoretical framework whereby the processes that shape human development may be examined and discovered. Bronfenbrenner expanded on the theoretical writings of other human developmental theorists before him, namely Lewin, Thomas and Thomas, Mead, and Freud.

Development of Five Ecological Systems

Bronfenbrenner sought to craft a view of human development that could explain growth without examining deficits within the person, which had been common practice in many developmental models previously developed. Before the introduction of Ecological Systems theory, **a General Systems theory was introduced by Bertalanffy.** In describing General Systems theory, Bertalanffy wrote the theory had moved from a primarily engineering and computer technology theory to a social sciences theory as a way to describe how humans interact with each other. Prior to the development of General Systems theory, most scientific study sought to reduce systems to the smallest units of measure and investigate them independently of each other; however, General Systems theory posited that whole systems could (and should) be examined intact with their interaction with each other being the important function for science to explore . While the development of General Systems theory pertained mainly to scientific and mathematical concepts at the time, the model may be applied within social science as its central role is to "**explain phenomena or order in terms of interactions of processes**"

Many social science studies from the 1950s through 1970s focused only on a child or only on the parents; the studies did not consider the reciprocal influences that the child and parent would have on each other. Bronfenbrenner proposed that systems of influence within an individual's immediate life each impacted the individual in different ways; however, he proposed the individual also had an impact on these systems. This whole-person approach in examining the individual within the environmental systems of influence is rooted in the concepts of General Systems theory. The systems within a person's environment occur at different levels, described by Germain as "Chinese boxes fit inside one another". Bronfenbrenner would propose the following systems: micro, meso, exo, and macro, later adding the chrono . **The micro, meso, and macro systems had been previously labelled by Brim.**

Microsystem

The person's immediate environment comprises a system of influence called the microsystem. Bronfenbrenner defined the microsystem as "a pattern of activities, roles, and interpersonal relations experienced over time by the developing person in a given setting with particular physical and material characteristics". Later, Bronfenbrenner would expand upon this definition: interaction in this level may be social or symbolic and should be sustained in increasingly complex ways , and the microsystem includes an individual's personality, beliefs, and temperament (1989). People with whom an individual has daily, face-to-face contact such as families or other people living in the home are also a part of the microsystem. The microsystem may include people outside of the home as well if they have regular, consistent contact with that individual. For children, this may be school or a daycare; for adults, this may be a workplace, gym, or coffee shop.

According to Bronfenbrenner, the importance of this level of the ecosystem cannot be understated. Bronfenbrenner and Evans wrote that the microsystem provides the outline for the processes that will influence the psychological development and behavioral changes of the individual. Experiences are critical to understanding how the influences of the microsystem work. Direct experience through contact within microsystems has a greater influence than indirect contact through or with other systems . Importantly, Bronfenbrenner noted that the systems that comprise the microsystem may affect the individual separately and in tandem with each other. This distinction leads to the second level of the Ecological System.

Mesosystem

The mesosystem describes the second level of the ecology of human development. Bronfenbrenner defined the mesosystem simply as “a system of microsystems". The mesosystem is made up of linkage between the different Microsystems in a person's life. In other words, how school and home interact creates a mesosystem. As noted above, Bronfenbrenner conceptualized different systems may work with(or against) each other in an individual’s life. These interactions between multiple microsystems create the mesosystem layers. Additionally, an important concept in the mesosystem development is that of synergy; that is, “the interaction of developmentally instigative or inhibitory features and processes[thatmaybe]present in each setting”.

School provides an example for this. The classroom, students in the class, and teacher create a microsystem for an individual student. The school itself, the student body, and the staff at the school comprise a mesosystem for an individual student. Microsystems and mesosystems must necessarily include the individual. Systems that impact one of these environments, but do not include the individual, create the third layer of the ecological environment.

Exosystem

The exosystem is much like the mesosystem in that it is made up of microsystems that interact with each other; however, in the exosystem, at least one of the microsystems cannot contain the person at the center of this system . A simple example is the workplace of a parent. The child is not a part of the workplace system; nevertheless, he or she could easily be influenced by that system if the parent is required to work long hours, possibly missing school events or even simply coming home stressed from work. Because the child is not a part of the workplace environment, this cannot be a part of his or her micro or mesosystems. This influence, then, occurs at the exosystem level.

The idea of reciprocity is easily seen in the lower systems. One can understand how a child may influence his or her parent and how that parent in turn influences the child. Even within the mesosystem, a child can easily influence both the parent and the teacher, while they are in turn influencing each other. This notion is still true within the exosystem as well. Even though the exosystems do not contain the individual, the influence is still reciprocal. Just as a parent may be influenced by the work system and come home stressed, a parent may also be stressed within the home system and bring this to work. A sick child may cause a parent to miss work, there by impacting the work system without the child being a part of it.

Macrosystem

The macrosystem is defined broadly as the overall culture and societal structure. The macrosystem creates a pattern of interaction between and among the different micro-, meso-, and exosystems. "The macrosystem may be thought of as societal blueprint for a particular culture or subculture" . Bronfenbrenner described the ways that the culture of a family develops within the structure of the family in the microsystem. This then is influenced by the mesosystems and exosystems of the individuals within that family. Furthermore, all of these systems are then impacted by the overall society and culture. Special emphasis is given to cultures within the groups, opportunity structures that are created by systems and experienced by individuals, and patterns of exchange within and among groups In later works, Bronfenbrenner added that the pattern of **social exchange** is an additional important component of the macrosystem.

Consider an immigrant family with a culture that is distinct from the majority culture. They may speak a language other than the majority at home. The adults may not speak the majority language at all while the children, because of school and socialization within their micro-and mesosystems, may speak both their native language and the majority language. While these influences are occurring at multiple levels of the individual systems, the macrosystem impact oftentimes may drive these forces which may vary depending upon the situation, the time, and the place in which they occur. This may then include the final level of the ecological perspective.

Chronosystem

When Bronfenbrenner first developed the Ecological Systems theory, he did not account for time as a construct in human development. Later, he added this concept, noting that many human developmental theorists had only looked at time as it pertained to the process of aging . That is, as people age and grows older, they mature and develop as their biological conditions changes. The addition of the chronosystem incorporates the concept of time into the ecological system of human development .This includes not only the aging and maturation of the person but also the time in which that person lives and develops. An example of this provided by Bronfenbrenner is a study of children and adolescents who grew up during the start of the Great Depression being affected differently than children born just after it. An example from more modern times might be children born pre and post 9/11 in the United States having vastly different concepts of terrorism and safety. The effects of the chronosystem do not have to be from a major event though. Consideration is given to the time and place in which the person lives. A young adult living in 21st century America would have a much different idea of privacy and its impacts than a young person living in a Brazilian rain forest in the mid-19th century for instance. Important aspects of the theory to consider.

Ecological transitions are important to the development of human ecology. Bronfenbrenner defined an ecological transition as the movement within a microsystem that changes or alters the makeup of that microsystem. A promotion at work, changing grades at school, the birth of a new sibling, or a death of a family member could all be examples of the alteration of a microsystem that change the makeup of that system. Bronfenbrenner identified that magnitude of these types of transitions in influencing the development of the individual stating that "**setting transitions continue to have developmental impact throughout the lifespan**" .

Examples

Systems	Examples
Microsystem	• Immediate family (parents, siblings, grandparents) • School environment (teachers, classmates) • Peer group and close friends • Extracurricular activities (sports teams, clubs) • Healthcare providers (pediatrician, dentist) • Neighborhood playmates • Childcare arrangements
Mesosystem	• Parent-teacher communication • Family-peer group interactions • School-neighborhood connections • Family-healthcare provider relationships • Interactions between different friend groups • Family-extracurricular activity connections • Religious community-family interactions
Exosystem	• Parents' workplaces and policies • Extended family networks • Local community organizations • School board decisions • Social services and support systems • Mass media and social media • Local government policies • Public transportation systems
Macrosystem	• Cultural norms and expectations • Socioeconomic factors • Educational policies and standards • Healthcare systems • Technological advancements • Environmental attitudes and policies • Gender roles and expectations • Religious or philosophical ideologies

Chronosystem	• Major historical events (e.g., pandemics, wars) • Technological shifts (e.g., rise of internet, social media) • Changes in family structure (e.g., divorce, remarriage) • Educational reforms • Economic cycles (booms and recessions) • Climate change and environmental shifts • Generational cultural changes • Personal life transitions (e.g., puberty, starting school)

1.3 Developmental ecology through space and time: process-person-context-time (PPCT) model

Evolution & Relevance of Bronfenbrenner's Theory in the 21st Century

Bronfenbrenner's theory of human development has undergone significant evolution since its inception in the 1970s, raising questions about its current relevance and application. Initially conceptualized as an ecological model focused primarily on contextual influences, it matured into a more sophisticated **bioecological model** emphasizing the critical role of proximal processes in development.

The mature version of the theory, often referred to as the bioecological model, places proximal processes at its core.

These processes are defined as "**enduring forms of interaction in the immediate environment**" and are considered the primary engines of development.

Central to the mature theory is the Process-Person-Context-Time (PPCT) model. This model emphasizes the interplay between four key elements:

1. **Process**: The core proximal processes driving development
2. **Person**: Individual characteristics that influence these processes
3. **Context**: The environmental systems in which development occurs
4. **Time**: The temporal aspect of development, including both individual life course and historical time

Despite these advancements, the theory's relevance in the 21st century has been a subject of debate. Kelly and Coughlan (2019) found significant links between Bronfenbrenner's ecological systems theory and contemporary frameworks for youth mental health recovery. Their research suggests that the components of mental health recovery are embedded in an "ecological context of influential relationships," aligning with Bronfenbrenner's emphasis on the importance of interconnected environmental systems. However, the rapid technological advancements of the 21st century have raised questions about how well Bronfenbrenner's theory accommodates these changes.

The theory's relevance is further challenged by common misapplications in contemporary research. Many scholars continue to apply outdated versions or misinterpret key concepts when claiming to use Bronfenbrenner's theory, as pointed out by other scholars. These misapplications often involve focusing solely on contextual influences without considering proximal processes, or failing to account for the time dimension in research designs.

Despite these challenges, Bronfenbrenner's theory remains a valuable framework for understanding human development in the 21st century. Its comprehensive nature allows for the examination of development in various contexts and across different life stages. The theory's emphasis on the interplay between individual characteristics, environmental influences, and temporal factors provides a nuanced approach to understanding the complexities of modern human development. To maintain its relevance, researchers and practitioners must understand the theory's evolution and apply it correctly. This involves recognizing the centrality of proximal processes, considering the role of technology in developmental contexts, and designing studies that capture the dynamic nature of development over time. By adapting the theory to include modern contexts while maintaining its core principles, Bronfenbrenner's bioecological model can continue to provide valuable insights into human development in the 21st century and beyond.

The bioecological shift highlighted reciprocal processes between the actively evolving individual and their immediate settings. This represented an evolution in Bronfenbrenner's thinking toward a more dynamic developmental process view. However, the bioecological model still acknowledged the broader environmental systems from his original theory as an important contextual influence on proximal processes.

The bioecological focus on evolving person-environment interactions built upon the foundation of his ecological systems theory while bringing developmental processes to the forefront.

1.4 An Integrated model of Individual Development based on PPCT Model

Neo-ecological theory

Navarro & Tudge (2022) proposed the neo-ecological theory, an adaptation of the bioecological theory. Below are their main ideas for updating Bronfenbrenner's theory to the technological age:

1. Virtual microsystems should be added as a new type of microsystem to account for online interactions and activities. Virtual microsystems have unique features compared to physical microsystems, like availability, publicness, and asychnronicity.

2. The macrosystem (cultural beliefs, values) is an important influence, as digital technology has enabled youth to participate more in creating youth culture and norms.
3. Proximal processes, the engines of development, can now happen through complex interactions with both people and objects/symbols online. So, proximal processes in virtual microsystems need to be considered.

Classroom Application

The Ecological Systems Theory has been used to link psychological and educational theory to early educational curriculums and practice. The developing child is at the center of the theory, and all that occurs within and between the five ecological systems is done to benefit the child in the classroom.

- According to the theory, teachers and parents should maintain good communication with each other and work together to benefit the child and strengthen the development of the ecological systems in educational practice.
- Teachers should also understand the situations their students' families may be experiencing, including social and economic factors that are part of the various systems.
- According to the theory, if parents and teachers have a good relationship, this should positively shape the child's development.
- Likewise, the child must be active in their learning, both academically and socially. They must collaborate with their peers and participate in meaningful learning experiences to enable positive development.
- There are lots of studies that have investigated the effects of the school environment on students. Below are some examples:
- Lippard *et al.* (2017) conducted a study to test Bronfenbrenner's theory. They investigated the teacher-child relationships through teacher reports and classroom observations.
- They found that these relationships were significantly related to children's academic achievement and classroom behavior, suggesting that these relationships are important for children's development and supports the Ecological Systems Theory.
- Wilson *et al.* (2002) found that creating a positive school environment through a school ethos valuing diversity has a positive effect on students' relationships within the school. Incorporating this kind of school ethos influences those within the developing child's ecological systems.

- Langford *et al.* (2014) found that whole-school approaches to the health curriculum can positively improve educational achievement and student well-being. Thus, the development of the students is being affected by the microsystems.

2

The Physical Ecology and Human Development

2.1 Developmental tasks through lifespan -role of ecology, social roles and structure role of ecology

Lifespan development is a progressive process of development in a human being involving an increase in age, which begins at conception and ends with death (Sugarman, 2000, p. 56). In addition, lifespan development can be divided into four levels depicting advanced functionality and character changes as an individual moves from one level to another. These levels include childhood and adolescence; early adulthood; middle adulthood; and late adulthood (Sugarman, 2000, p. 56).

Concept of Developmental task

The developmental task concept occupies middle ground between two opposed theories of education: the **theory of freedom**—that the child will develop best if left as free as possible, and the **theory of constraint**—that the child must learn to become a worthy, responsible adult through restraints imposed by his society. A developmental task is midway between an individual need and societal demand. It assumes an active learner interacting with an active social environment. Mastery of these tasks is satisfying and encourages us to go on to new challenges. Difficulty with them slows progress toward future accomplishments and goals. As a mechanism for understanding the changes that occur during the life span. Robert Havighurst (1952, 1972, 1982) has identified critical developmental tasks that occur throughout the life span. Although our interpretations of these tasks naturally change over the years and with new research findings. Havighurst's developmental tasks offer lasting testimony to the belief that we continue to develop throughout our lives. Havinghurst (1972) defines a developmental task as one that arises at a certain period in our lives, the successful achievement of which leads to happiness and success with later tasks; while leads to unhappiness, social disapproval, and difficulty with later tasks.

Characteristics of Developmental Task

Havighurst considered the many different aspects of a person's life that influence a person's development viz., the biological development and physical structures of the individual; the society in which the person lives, and the resultant cultural influences, as well as the individual's personal characteristics, values and goals. This view of development takes into account the role of physical maturation and the role that society plays in determining the skills that need to be learned at a certain age.

According to Havighurst, there are sensitive periods which he called teachable moments, when an individual is mature enough to learn the developmental tasks. These tasks may be physical like walking, cognitive like learning to read, or social where the person develops significant relationships. Once the critical period of development is over, learning may still occur. Language skills for example, continue to develop as one learns more complex ways of using language. Some tasks are the same for everyone, regardless of where you live, as they are about human biology. Some tasks are different in different societies. Havighurst added that biology and society have a lot influence, but personal values of the people who prepare them have as much influences on identifying the task.

Sources of developmental tasks

Developmental tasks arise from three different sources (Havighurst, 1948, 1953). In his bio psychosocial model, the first important issue is biology, second is psychology and the last one is the sociology. He identifies three sources of developmental tasks (Havighurst, 1972)

- First, some tasks are mainly based on physical maturation. Tasks that arise from personal sources.
- Another source of developmental tasks involves personal values and aspirations. These personal factors result from the interaction between ontogenetic and environmental factors, and play an active role in the emergence of specific developmental tasks (example, choosing a certain occupational pathway).
- The third source of developmental tasks relates to socio-structural and cultural forces-i.e., tasks that have their source in the pressures of society. Such influences are based on, for instance, laws (example, minimum age for marriage) and culturally shared expectations of development. These tasks may be physical like walking, cognitive like learning to read, or social where the person develops significant relationships or learning the role of a responsible citizen. .

Once the critical period of development is over, learning may still occur. Language skills for example, continue to develop as one learns more complex ways of using language. Havighurst believed that human development is a process in which people attempt to learn the tasks required of them by the society to which they are adapting.

Sensitive Period

It is the period of time when the person is most ready to acquire a new ability. Mastery of a developmental task is influenced by the resolution of the psychosocial crisis of the previous stage and this resolution leads to the development of new social capabilities. These capabilities orient a person towards new experiences, a new aptitude for relationships and new feelings of personal worth as he/she confronts the challenges of developmental tasks of the next stage.

Psychological crisis: a normal set of stresses and strains arise, because one must make psychological efforts to adjust to the demands of social environment at each stage of development. Societal demands vary from stage to stage. People experience these demands as mild but persistent guidelines for and expectations of behaviour. Before the end of each stage of development, the individual tries to achieve a resolution to adjust to societal demands and at the same time to translate those demands into personal gains. This process produces a state of tension that the individual must reduce in order to proceed to the next stage.

The developmental tasks concept has a long and rich tradition. Its acceptance has been partly due to a recognition of sensitive periods in our lives and partly due to the practical nature of Havighurst's tasks. Havighurst is not alone in the importance he places on the developmental task concept (Cole, 1986; Goetting, 1986; Cristante & Lucca, 1987; Cangemi and Kowalski, 1987). For example, Goetting (1986) has examined the developmental tasks of siblings and identified those that last a lifetime, such as companionship and emotional support. Other tasks seem to be related to a particular stage in the life cycle, such as caretaking during childhood and later the care of elderly parents.

Developmental tasks in each stage

Identifying and mastering developmental tasks help us to understand the way change affects our lives. Another way to understand life span changes is to identify those needs that must be satisfied if personal goals are to be achieved. To recognize the role that needs play in our lives, it is important to examine the work of Abraham Maslow and his needs hierarchy.

Developmental tasks of Infancy and early childhood: 1.Learning to walk 2. Learning to take solid foods 3. Learning to talk 4. Learning to control the elimination of body wastes 5. Forming concepts and learning language to describe social and physical reality. 6. Getting ready to read

Ages from birth to 6-12 yrs (Middle childhood) 1.Learning physical skills necessary for ordinary games. 2. Building wholesome attitudes toward oneself as a growing organism 3. Learning to get along with age-mates 4. Learning an appropriate masculine or feminine social role 5. Developing fundamental skills in reading, writing, and calculating 6. Developing concepts necessary for everyday living. 7. Developing conscience, morality, and a scale of values 8. Achieving personal independence 9. Developing attitudes toward social groups and institutions

Developmental tasks of adolescence(12-18 yr): 1.Achieving new and more mature relations with age mates of both sexes 2. Achieving a masculine or feminine social role 3. Accepting one's physique and using the body effectively 4. Achieving emotional independence of parents and other adults 5. Preparing for marriage and family life preparing for an economic career 6. Acquiring a set of values and an ethical system as a guide to behaviour; developing an ideology 7. Desiring and achieving socially responsible behaviour

Developmental tasks of early adulthood: 1. Selecting a mate 2. Achieving a masculine or feminine social role 3. Learning to live with a marriage partner 4. Starting a family 5. Rearing children 6. Managing a home 7. Getting started in an occupation 8. Taking on civic responsibility 9. Finding a congenial social group

Adulthood Developmental Tasks (19-30 years): 1. Selecting a mate 2. Learning to live with a marriage partner 3. Starting a family 4. Rearing children 5. Managing a home 6. Getting started in an occupation 7. Taking on civic responsibility 8. Finding a congenial social group

Middle-age Developmental Tasks (30 – 60 years): 1. Achieving adult civic and social responsibility 2. Establishing and maintaining an economic standard of living 3. Assisting teen-age children to become responsible and happy adults 4. Developing adult leisure, time activities 5. Relating oneself to one's spouse as a person 6. Accepting and adjusting to the physiological changes of middle age7.Adjusting to ageing parents.

Old Age or Later Maturity Developmental Tasks (60 Years and Over): Havighurst forward the following developmental tasks for this view. 1. Adjusting to decreasing physical strength and health: 2. Adjusting to retirement and reduced income 3.Adjusting to death of spouse 4. Establishing an explicit affiliation with one's age group 5. Meeting social and civic obligations 6. Establishing satisfactory physical living

Havighurst has divided these tasks into four major categories

1. **Tasks relating to physical changes:** These include the acceptance of, and adjustment to physical changes that normally occur during middle age.
2. **Tasks relating to changed interests:** The middle aged person often assumes civic and social responsibilities and develops an interest in adult oriented leisure time activities which were not taken care of during early adulthood.
3. **Tasks relating to vocational adjustments:** These tasks revolve around establishing and maintaining a relatively stable standard of living.
4. **Tasks relating to family life**: The major tasks include relating to oneself to one's spouse as a person, adjusting to aging parents, and assisting teen age children to become responsible and successful adults.

The lifespan perspective examines the changes that take place at each level of human development relative to the environmental (Society and culture) factors that influence these changes(Sigelman & Rider, 2008). Moreover, the lifespan perspective can be defined in many ways. For instance, the lifespan perspective can be a progressive and lifelong process of development that is not limited to any single level in human development. On the other hand, there are three major aspects of the lifespan perspective of development, which include the cognitive, social, physical dimensions of change (Sugarman, 2000, p. 59).

Furthermore, other studies claim that the perspective is plastic in nature because some domains of lifespan development increase while others decrease. This plasticity of the lifespan perspective occurs in response to various environmental factors that influence human development. In addition, the perspective is embedded in the lifelong events that occur in the life of an individual. Thus, this perspective is studied by scientists, anthropologists, and psychologists, which brings out its multidisciplinary nature (Sugarman, 2000). On the other hand, the perspective has different contextual implications. Therefore, human development is determined by biological, cultural, social, and physical environmental factors.

The three aspects of the lifespan perspective are also implicated in different developmental changes. The physical aspect of development involves changes in weight, height, shape, and the changes in individual experiences with the external environment (Sigelman & Rider, 2008).

There are three great groups of phenomena in the global system of the noosphere (The noosphere represents the highest stage of biospheric development, that of humankind's rational activities as viewed by biogeochemist) which qualitatively differ from one another: phenomena related to the structure of the non-organic matter, biological and social phenomena, but they are primarily related to man. The evolutional processes that are characteristic of the functioning of the noosphere specifically occur in each of these groups of phenomena, and during the transition from one group into another. In the multitude of revolutionary leaps in the history of the world each leap has enormous importance as it plays a respective crucial role in the evolution of man on the Earth. The first transition is from **non-organic to organic matter** when life on the planet is created. The second one is connected with the **origin of cognition and appearance of man**. Both these transitions are based upon a number of definite **quantitative changes that lead to the qualitative leap** and the appearance of a new type of relations.

Explaining Lifespan Development with theoretical perspective

There are many theories that attempt to explain the process of human development from different perspectives such as Freud's theory of psychosexual development and Erikson's theory on psychosocial development. Erikson's theory posits that an individual encounters several conflicts during the development process, which depend on the type of relationship existing between this individual and the society.

Therefore, as an individual moves through the eight stages of development discussed above, he/she must solve the conflicts involved in one stage before moving to another. This ensures that one develops a sound personality; otherwise one may encounter difficulties in addressing conflicts in subsequent stages if the previous ones were not sufficiently resolved (Sugarman, 2000).

Freud's theory states that the process of personality development occurs at the early childhood stage. Subsequently, the behavioral changes observed in developing individuals are influenced by the childhood events. Here, personality development occurs through several stages during the early childhood stage. During this stage, the pleasure-oriented capabilities of an individual become focused on specific areas in one's body. These capabilities also known as sexual libido or psychosexual energy play a major role in the subsequent behavioral changes in an individual (Sigelman & Rider, 2008). Thus, if all the issues involved in the psychosexual stages of development are adequately resolved, then an individual develops a sound personality. However, if the psychosexual issues remain unresolved, the person involved is trapped in a particular stage until all the issues are resolved.

On the other hand, the social aspect of development involves changes in different phases of human development, which are directly influenced by the social environment such as social skills and relationships. Lastly, the cognitive domain of human development entails change in an individual's thinking capacity, memory, and decision-making.

The changes characterizing the lifespan perspective of development can also be divided into eight stages of human development. Erik Erikson was an ego psychologist who developed one of the most popular and influential theories of development. While his theory was impacted by psychoanalyst Sigmund Freud's work, Erikson's theory centered on psychosocial development rather than psychosexual development. The **first stage** also known as the **Trust vs. Mistrust** stage occurs during the period between birth and one year.

Here, development of trust depends on the relationship between the toddler and the care-giver (Sugarman, 2000, p. 66). The **second stage** also known as **Autonomy vs. Shame** and Doubt period occurs during the early childhood stage. During this stage, the child is capable of making choices relative to the individualized willingness.

The **Initiative vs. Guilt is the third stage** in development, which occurs during the middle childhood stage. During this stage, children develop a sense of purpose through engaging in goal-oriented activities (Sigelman & Rider, 2008, p. 88). The **fourth stage also called the Industry vs. Inferiority** period occurs at the late childhood period. Here, children develop competence through learning social norms, basic education, and culture.

Additionally, the adolescence stage signifies the **Identity vs. Role** Confusion period. Here, individual values and choices develop because individuals are able to account for their actions. The **Intimacy vs. Isolation** period occurs during early adulthood and it is characterized by development of intimate relationships, marriage, and families (Sigelman & Rider, 2008, p. 90).

The **Generativity vs. Stagnation** period occurs during the middle adulthood stage. Here, individuals are committed with providing for their families and developing their careers. Lastly, the **Integrity vs. Despair** stage occurs during the late adulthood stage. The elderly individuals are self-contented and they are full of life experiences and advice.

Social ecology is the study of how individuals interact with and respond to the environment around them, and how these interactions affect society and the environment as a whole. Consider traditional ecological studies, in which students examine how multiple factors must interact within nature to create the world's ecosystems.

Social ecology takes the same approach by examining society holistically Studying how individuals, collectives and institutions interact with and depend on one another, social ecologists look at the bigger picture of our "system"—allowing for a more effective approach to solving society's collective problems.

What is Social Ecology?

The concept of social ecology was introduced by an environmental activist named Murray Bookchin. He was an ecologist and believed there was a better approach to the study. In his paper "What is Social Ecology?" he argues environmentalists are too focused on studying the individual symptoms of a problem rather than addressing the problem itself—the belief that humans can and should control nature.

This philosophy served as the foundation for social ecology. In Bookchin's eyes, the hierarchies that exist within society are to its own detriment. The issues and dysfunctions of human society, whether they are environmental or social, come from a human-made hierarchical structure. Rather than a hierarchy, Bookchin believed that life and society should be looked at as an ecosystem where all the moving parts are equally important to a healthy, stable and sustainable environment.

The belief is that following these principles and extending them to all facets of society will lead to a more equal, cooperative environment where hierarchies don't decide winners and losers. Recognizing the importance of each 'piece' to the system as a whole is a fundamental part of a good social worker's mindset.

The social ecological perspective points out that many factors from our environment come together to create the unique circumstances that shape who we are, seeking to promote the idea that we are all interdependent and must handle society's issues in ways that consider all parts of a functional system. Applying this ecological perspective in social work can benefit your existing understanding. Often, those in need of social assistance are those at the bottom of the hierarchy, such as deeply impoverished families.

Analytical Framework for Understanding the Social-Ecological System (SES)

Some of the researchers have referred to SES as an integrated system of ecosystems and human society with reciprocal feedback and interdependence, consisting of multiple subsystems and internal variables (Folke *et al* 2010; Ostrom, 2009). So,in a broader term, SES tries to explain the interdependence or the reciprocity between nature and society, regarding humankind as the key subject. The concept of SES has been used by several researchers to analyse the system of resource management (Berkes et al 2000; Berkes *et al* 2003; Ostrom, 2009). They tried to distinguish between the westernized concept of resource management i.e. conservation with the local resource management system and argue the latter is more sustainable and thus should be desired. Colding and Barthel (2019)reviewed 20 years' research papers on social-ecological systems and concluded that there is a lack of comprehensive analytical frameworks for understanding or analyzing the SES framework. However, there are a few frameworks which have been used to understand the concept of SES, namely Holing's concept of 'adaptive cycle' which includes four stages i.e. Exploitation, Conservation, Release and Reorganisation (Holling and Gunderson, 2002) and Ostrom's general framework for analyzing the sustainability of SESs (Ostrom, 2009). As the present paper aims to study the structure and functioning of the social-ecological system of the area, Ostrom's framework of SES is being used.

According to Ostrom (2009), there are many subsystems and internal variables inside a larger and complex SES, which describes the inter linkages of resource management which vary across space and time. The following figure includes the first level four core systems i.e. Resource Systems (RS),Resource Units(RU), Governance Systems(GS)and Users(U).

Social role of ecology in human development perspective

The social role of ecology is to study how humans interact with their environment, analyzing how these interactions impact society and the environment as a whole, essentially highlighting the interconnectedness between human social structures and the natural world, and informing solutions to environmental problems through understanding the social factors involved in environmental decision-making and behavior

Key points about the social role of ecology

- **Understanding social impacts**: Ecology helps identify how environmental changes can affect social dynamics, such as access to resources, food security, migration patterns, and community health.
- **Policy development**: By understanding the social dimensions of environmental issues, ecologists can inform policy decisions to promote sustainable practices and address environmental concerns with community input.
- **Community engagement**: Social ecology encourages community involvement in environmental decision-making, considering local knowledge and perspectives when addressing ecological issues.
- **Social justice considerations**: It highlights how environmental issues disproportionately affect marginalized communities, promoting equitable solutions.
- **Behavioral change**: By understanding social norms and motivations, ecology can inform strategies to encourage pro-environmental behaviors at the individual and community level.

- **By understanding social norms and motivations,** ecology can inform strategies to encourage pro-environmental behaviors at the individual and community level.

The application of social ecological theories and models focus on several goals: (i) to explain the person-environment interaction, (ii) to improve people-environment transactions, (iii) to nurture human growth and development in particular environments, and (iv) to improve environments so they support expression of individual's system's dispositions. Some examples are:

Political and economic policies that support the importance of parent's roles in their children's development such as Head Start or Women Infants and Children programs. Fostering of societal attitudes that value work done on behalf of children at all levels: parents, teachers, extended family, mentors, work supervisors, legislators.

In community health promotion: identifying high impact leverage points and intermediaries within organizations that can facilitate the successful implementation of health promoting interventions, combining person focused and environmentally based components within comprehensive health promotion programs, and measuring the scope and sustainability of intervention outcomes over prolonged periods. Basis of intervention programs to address issues such as bullying, obesity, overeating and physical activity. Interventions that use the social ecological model as a framework include mass media campaigns, social marketing, and skills development.

In economics: economics, human habits, and cultural characteristics are shaped by geography. In economics, an output is a function of natural resources, human resources, capital resources, and technology. The environment (macrosystem) dictates a considerable amount to the lifestyle of the individual and the economy of the country. For instance, if the region is mountainous or arid and there is little land for agriculture, the country typically will not prosper as much as another country that has greater resources.

In risk communication: used to assist the researcher to analyze the timing of when information is received and identify the receivers and stakeholders. This situation is an environmental influence that may be very far reaching. The individual's education level, understanding, and affluence may dictate what information he or she receives and processes and through which medium.

Education and Awareness: Ecology fosters a deeper understanding of the natural world and our place within it. Educational programs and initiatives that emphasize ecological principles can inspire individuals to adopt more environmentally friendly behaviors and advocate for policies that protect the environment.

There are three great groups of phenomena in the global system of the noosphere (The noosphere represents the highest stage of biospheric development, that of humankind's rational activities as viewed by biogeochemist) which qualitatively differ from one another: phenomena related to the structure of the non-organic matter, biological and social phenomena, but they are primarily related to man. The evolutional processes that are characteristic of the functioning of the noosphere specifically occur in each of these groups of phenomena, and during the transition from one group into another. In the multitude of revolutionary leaps in the history of the world each leap has enormous importance as it plays a respective crucial role in the evolution of man on the Earth. The first transition is from non-organic to organic matter when life on the planet is created. The second one is connected with the origin of cognition and appearance of man. Both these transitions are based upon a number of definite quantitative changes that lead to the qualitative leap and the appearance of a new type of relations.

Climate Change Mitigation: Ecology provides essential insights into the impacts of climate change and the ways in which ecosystems can help mitigate its effects. For instance, forests and oceans act as carbon sinks, sequestering carbon dioxide from the atmosphere. Understanding these dynamics informs efforts to combat climate change and adapt to its impacts.

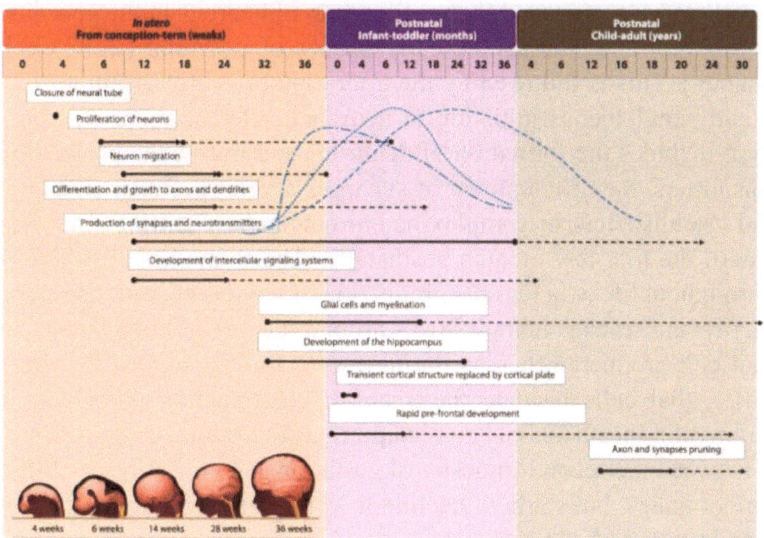

Depicts approximate timelines for experience-dependent synaptic development:

Visual and auditory cortex (seeing/hearing); Angular gyrus/Broca's area (language and speech);

Prefrontal cortex (higher cognitive functions)

Rapid development — Prolonged development

In personal health: to prevent illnesses, a person should avoid an environment in which they may be more susceptible to contracting a virus or where their immune system would be weakened.

In human nutrition: used as a model for nutrition research and interventions. The social ecological model looks at multiple levels of influence on specific health behaviors. Levels include intrapersonal (individual's knowledge, demographics, attitudes, values, skills, behavior, self-concept, self-esteem), interpersonal (social networks, social supports, families, work groups, peers, friends, neighbors), organizational, community and public policy level, Multi-level interventions are thought to be most effective in changing behavior. Access to healthy food supports physical and cognitive development. Research indicates that malnutrition can lead to developmental delays and learning difficulties.

Nutrition is critical in supporting healthy brain development early in life, with long-lasting, and often, irreversible effects on an individual's cognitive development and life-long mental health. In this review, we present recent human and pre-clinical evidence on the role of nutrition, with particular focus on more emerging nutrients, in neurocognitive development in healthy infants and children aged 0–59 months.

Development of the human brain starts with the closure of the neural tube by the fourth week of pregnancy and the proliferation of neurons in the germinal layers near the ventricles during the early phases of gestation (from week six of pregnancy). This is followed by the migration of neurons to their final destination and simultaneous initiation of neuronal differentiation. Neuronal differentiation includes the formation of dendrites and axons, the production of neurotransmitters, the development of synapses and intracellular signaling systems, and establishment of complex neural membranes starting from late pregnancy until the first few months postnatally. The formation of synapses continues throughout life ,whereas the production of various neurotransmitters starts prenatally and reaches mature levels around the age of three years . In parallel, glial cell production begins during the second trimester (32 weeks of gestation) . glial cells insulate the axons by surrounding them with a membranous myelin sheath (axonal myelination), a process that predominantly takes place between the second trimester of gestation and the first year of life. Myelination continues, but starts to decline in early adulthood after which it stops at around the age of 40 years.

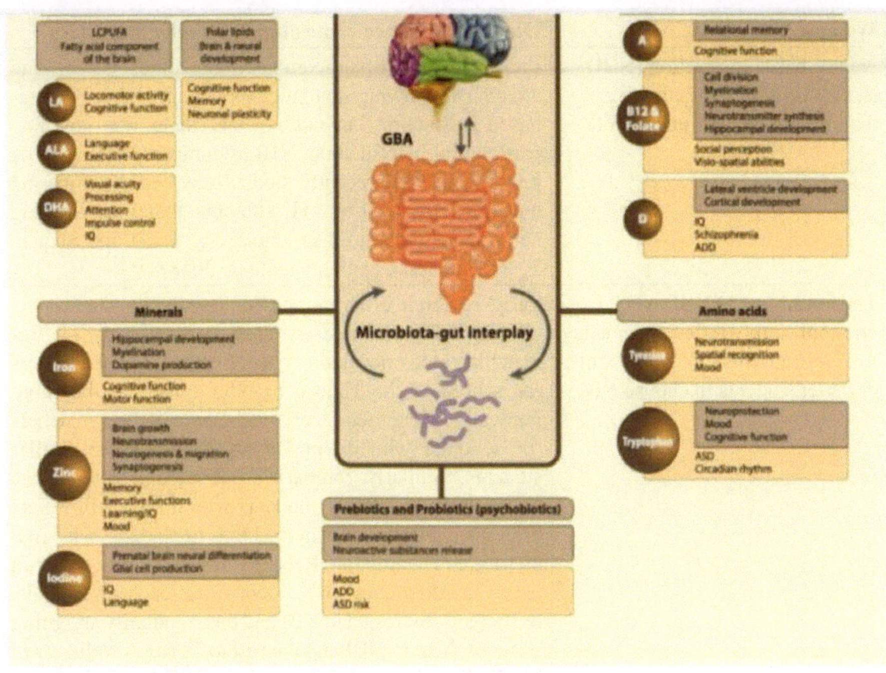

Roles in nervous system development	Affected domain if deficient

LCPUFA: long-chain polyunsaturated fatty acid; LA: linoleic acid; ALA: alpha-linolenic acid; DHA: docosahexanoic acid; IQ: intelligence quotient; ASD: autism spectrum disorder; ADD: attention deficit disorder; GBA: Gut-Brain Axis

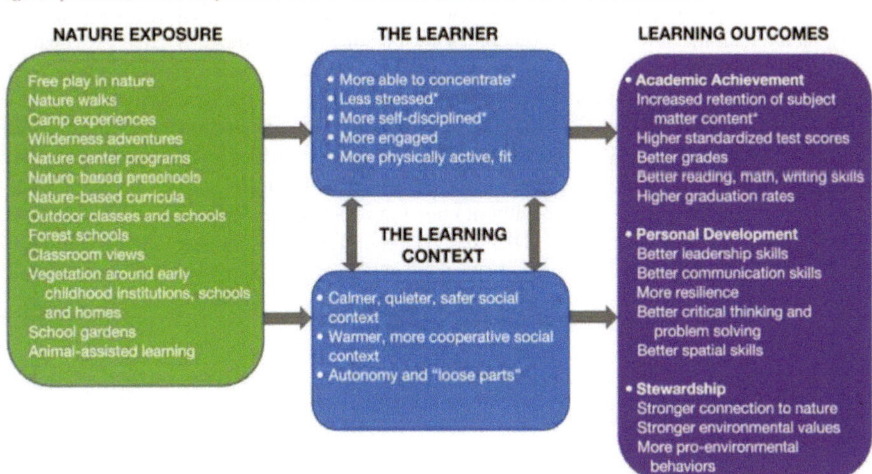

We now know that...	How this advance came about and why it matters
Nature-based instruction (NBI) is, on average, more effective than traditional instruction (TI).	Early research often compared outcomes before and after NBI, showing that students benefited from nature-based instruction but not whether there was anything particularly helpful about NBI as compared to any other instruction. More recently, studies have begun comparing outcomes for NBI vs. TI, showing that incorporating nature adds value to instruction (e.g., Emst and Stanek, 2006; Camassao and Jagannathan, 2018).
The advantage of NBI over TI does not simply reflect a tendency for better teachers, better schools, or better students to choose NBI.	Early research often compared learning in classrooms offering NBI vs. matched classrooms offering TI. But matching does not address the likelihood that teachers (or schools) who choose to offer NBI may be more innovative, energetic, or well-funded than teachers (or schools) who do not, even if they serve similar students. Similarly, comparisons of students who choose extracurricular NBI vs. students who do not will reflect pre-existing differences in the kinds of students who sign up for extra instruction. Recently, researchers have begun using "**waitlist controls**" - identifying teachers, schools, or students interested in NBI and then randomly assigning some of them to NBI and the rest to TI (e.g., Wells *et al.*, 2015) Guarding against pre-existing differences between the teachers, schools, and students being compared lends greater confidence that any gains are due to the instruction itself.
The effects of NBI on academic learning are real; they do not simply reflect the rosy assessments of biased observers	Early research often relied on subjective assessments of outcomes by persons who believe in NBI. Advocates, practitioners, and parents or children who choose NBI may perceive benefits in the absence of any real effects, whether consciously or unconsciously. More recent research guards against such bias by employing objective measures or assessments made "blind to condition-without knowing which students were in which condition (NBI or TI) (e.g., Emet and Stanek, 2006). In these studies, an advantage of NBI over TI cannot be attributed to wishful thinking.

NBI shows a "**dose-response relationship**" - as the magnitude of the treatment (the dose) increases, so does the outcome.	Early research relied on binary comparisons between learning settings with and without nature, or "low" and "high mature, leaving more room for alternative explanations. For instance, if students learn more outdoors than indoors, the difference might be due to either differences in vegetation or other differences between the settings. More recent research has compared multiple levels of nature (e.g., schoolyards with 0-40% tree cover, Sivarajah *et al.*, 2018) or multiple levels of NBI (Wells *et al.*, 2015). When the response is proportional to the dose, that lends confidence that the effect is attributable to the level of vegetation. Although a "dose-response relationship" does not prove causality, it strengthens the case.
The nature-leaming connection holds up across topics, learners, instructors, pedagogies, places, and measures of learning.	As researchers have continued to conduct studies, the body of studies testing the nature-learning hypothesis has grown larger and more diverse (e.g., Faber Taylor *et al.*, 2002; Maynard *et al.*, 2016: O'Haire *et al.*, 2013; Ruiz-Gallardo *et al.*. 2013; Fremery and Bogner, 2014; Lekies *et al.*, 2015; Swank *et al.*, 2017; ko *et al.*, 2018a; McCree *et al.*, 2018; Sivarajah *et al.*, 2018). A robust association persisting across different contexts lends greater confidence in a cause-and-effect relationship (Ha, 1965, p. 8).
The relationship between nature and learning holds up across different research designs.	Over time, a greater variety of study designs have been employed, including true experiments (e.g., Wells *et al.*, 2015), quasi-experiments (e.g., Faber Taylor and Kuo, 2009; Benfield *et al.*, 2010), large-scale correlational studies with statistical controls (e.g., Kuo and Faber Taylor, 2004), and longitudinal studies (e.g., McCree et al 2018). Findings persisting across diverse study designs strengthen the case for causality.
The advantages of NBI over TI may stem from both setting and pedagogy	Previous reviews drew only upon studies examining the effects of NBI on learning. In this review, we expanded our reach to include research on both the setting and the pedagogy of NBI, respectively. Educational psychologists working in the classroom have found that **active, hands-on, student-centered, and collaborative forms of instruction outperform more traditional instructional approaches** (Granger *et al.*, 2012; Freeman *et al.*, 2014; Kontra *et al.*, 2015). Environmental psychologists have found better learning in "greener" settings - even when the instruction does not incorporate the nature (Benfield *et al.*, 2015; Kuo *et al.*, 2018b). These additional bodies of evidence converge to support and perhaps explain the advantages of NBI over TI.

Nature experiences may promote learning via at least eight distinct pathways.	Again, previous reviews drew only upon direct tests of the nature-learning hypothesis - studies in which nature was the independent variable and learning was the dependent variable. In this review, we also examined studies in which nature was the independent variable but the dependent variable was a precursor to learning (for example, Li and Sullivan, 2016, examines impacts of classroom views of nature on attention, which has long been established as an important precursor to leaming, e.g., Rowe and Rowe, 1992). Evidence of mechanism lends greater plausibility to a cause-and-effect relationship between nature and learning.
	The multiple mechanisms identified here may also help explain the consistency of the nature-learning relationship, as robust phenomena are often multiply determined.
In recent years, the evidence for a cause-and-effect relationship between nature experiences and learning has advanced considerably. Some advances can be traced to the adoption of more rigorous research methods in individual studies (in green), others can be traced to the maturation of the field (in blue), and still, others stem from broadening the kinds of evidence considered in reviews (in purple).	

In public health: drawing upon this model to address the health of a nation's population is viewed as critically important to the strategic alignment of policy and services across the continuum of population health needs, including the design of effective health promotion and disease prevention and control strategies.

Environmental Quality: Clean air, water, and safe living conditions are crucial. Exposure to pollutants can affect cognitive development and health. Studies show that children living in areas with high pollution levels are more likely to have respiratory issues and cognitive impairments.

Cultural and Recreational Value: Natural landscapes and wildlife have profound cultural and recreational importance. They inspire art, literature, and traditions and offer spaces for recreation and tourism. Ecological awareness helps preserve these invaluable aspects of our cultural heritage.

Access to Green Spaces: Interaction with nature enhances cognitive and emotional development. A study by Wells and Evans (2003) found that children with greater access to nature had lower levels of stress and better psychological well-being.

Structural Role of Ecology in Human Development

In human development research, "structural ecology" refers to the study of how the physical and social structures within an environment, like neighborhoods, communities, or institutions, influence and shape an individual's development trajectory, essentially examining how the "structure" of one's environment impacts their life outcomes, often through the lens of Urie Bronfenbrenner's Ecological Systems Theory; it highlights the interconnectedness of various environmental factors and how they interact to influence human development across different life stages. Key points about the role of structural ecology in human development research:

Examining environmental contexts: Structural ecology goes beyond just individual characteristics to analyze the broader environmental structures that individuals are embedded within, like socioeconomic status, access to healthcare, quality of education, neighborhood safety, and social networks. With regard to the role of structure in shaping human development, we specified the relation of individual differences in all of these psychological determinants to cultural and historical norms, mediated through the beliefs and behaviors of parents, teachers, peers, the mass media, and so on. In particular, we linked achievement-related beliefs, outcomes, and goals to interpretative systems such as causal attributions and other meaning-making beliefs linked to achievement-related activities and events, to the input of socializers (parents, teachers, peers, media, members of one's neighborhood and other social groups), to various social roles and other culturally and historically based beliefs about the nature of various task in a variety of achievement domains and the "appropriateness" of participation in such tasks, to self-perceptions and self-concept, and to one's perceptions of the task itself, and to the processes and consequences associated with identity formation. We hypothesized that each of these factors influences the expectations one holds for future success at the various achievement-related options and the subjective value one attaches to these various options. **These expectations and the value attached to the various options, in turn, are assumed to influence choice among these options.**

- **Nested systems**: Bronfenbrenner's ecological systems theory conceptualizes these structures as nested systems, including the microsystem (immediate family), mesosystem (interactions between microsystems), exosystem (wider social systems like workplace), and macrosystem (cultural values and laws).

A child inherits the genetic information of both parents through the information carriers known as genes. The genes are made up of DNA, which is found on chromosomes. On the other hand, the environment in the context of lifespan development refers to the total social and cultural factors surrounding a developing human being (Sigelman & Rider, 2008, p. 86). Therefore, lifespan development in children depends on two major factors; the genetic make-up of a child and the environmental factors. However, the two factors can play a mutual role in influencing the development of different individuals in the society. Here, the epigenetic framework of development posits that the genetic make-up can be turned on and off relative to the internal and external environmental feedback (Sigelman & Rider, 2008,). This gives the process of lifespan development the **flexibility and plasticity** described in the discussions above. For instance, the learning process in children follows a specific pattern of adaptability, which depends on the internal and external factors. Therefore, during the early childhood stage, individuals use the **environmental experiences** in reasoning and acting. However, as the environmental factors and experiences change along the period of lifespan development, so does the emotional, reasoning, and social attributes in an individual.

From the discussions above, the lifespan perspective of development examines the changes that occur in different individuals, which occur due to the interaction of the genetic make-up and the total circumstances surrounding an individual. On the other hand, this perspective is also reinforced by Freud's theory of psychosexual development and Erikson's theory of psychosocial development, which are summarized in the discussions above.

The role of physical ecological factors, terrain, climate changes, and demographics in influencing behavioral development in humans is multifaceted and complex. These factors shape human behavior through their impact on individuals' interactions with their environment, social relationships, cultural practices, and adaptive responses to environmental challenges. Let's explore each of these components in detail:

Studies examining interrelationship between different layers in ecological model by Bronfebrunner highlighting the social and structural role of ecology.

<table>
<tr><th>Systems</th><th>Examples</th><td rowspan="5">• Children with secure attachment are more likely to develop healthy relationships and exhibit resilience in the efface of challenges.
• The research by Bowlby (1969) on attachment theory underscores the importance of early relationships in shaping emotional and social development. Children who engage in meaningful interactions with family members, participate in educational activities and receive emotional support exhibit better academic performance and cognitive skills.
• Evans (2006) stated that, children living in cleaner, more organized homes exhibit better cognitive outcomes & fewer behavioural issues compared to those in cluttered environment.
• The National scientific council on the developing child highlights that positive early experiences and stable relationships with caregivers are foundational for healthy brain development , influencing emotional regulation and social skills.</td></tr>
<tr><td>Microsystem</td><td>• Immediate family (parents, siblings, grandparents)
• School environment (teachers, classmates)
• Peer group and close friends
•Extracurricularactivities(sportsteams, clubs)
• Healthcare providers (pediatrician, dentist)
• Neighborhood playmates
• Childcare arrangements</td></tr>
<tr><td>Mesosystem</td><td>• Parent-teacher communication
• Family-peer group interactions
• School-neighborhood connections
• Family-healthcare provider relationships
• Interactions between different friend groups
• Family-extracurricular activity connections
• Religious community-family interactions</td></tr>
<tr><td>Exosystem</td><td>• Parents' workplaces and policies
• Extended family networks
• Local community organizations
• School board decisions
• Social services and support systems
• Mass media and social media
• Local government policies
• Public transportation systems</td></tr>
<tr><td>Macrosystem</td><td>• Cultural norms and expectations
• Socioeconomic factors
• Educational policies and standards
• Healthcare systems
• Technological advancements
• Environmental attitudes and policies
• Gender roles and expectations
• Religious or philosophical ideologies</td></tr>
<tr><td>Chronosys-tem</td><td>• Major historical events (e.g., pandemics, wars)
• Technological shifts (rise of internet, social media)
• Changes in family structure (e.g., divorce, remarriage)
• Educational reforms
• Economic cycles (booms and recessions)
• Climate change and environmental shifts
• Generational cultural changes
• Personal life transitions (e.g., puberty, starting school)</td><td>• A meta analysis by goodman et al (2011) found that maternal depression is associated with increased risks of emotional and behavioural problems in children.</td></tr>
</table>

2.2 Physical ecological factors, Terrain, climatic changes, demographics and their effect on human development

Availability of Resources: The availability of resources such as food, water, shelter, and natural landscapes profoundly influences human behavior. In environments where resources are scarce, individuals may exhibit competitive behaviors to secure necessities for survival. Conversely, in resource-rich environments, cooperation and sharing may be more prevalent.

The availability of financial, human, information and physical resources are regarded as indispensable to achieve the desired goals in both personal and professional spheres and lead to up-gradation of one's living conditions. The individuals, belonging to all occupations, communities, categories and backgrounds need to put emphasis on availability of these resources throughout their lives. The financial resources are to be spent in an adequate manner to meet all the needs and requirements and to sustain one's living conditions in an adequate manner. Human resources make use of their skills and abilities to carry out all types of job duties in a well-organized manner. Information resources are the resources that make provision of information in terms of various features, subjects and concepts. Physical resources are the resources that are necessary to put into operation various types of job duties and tasks. Furthermore to sustain one's living conditions in an adequate manner, these resources are available. These include, infrastructure, amenities and facilities. The individuals need to be well-versed in terms of methods and approaches to manage and make use of resources in an effective manner. Therefore, it is well-understood that availability of resources is indispensable to enrich one's living conditions.

Access to Nature: Proximity to natural environments, such as forests, parks, and coastlines, can influence behavioral patterns. Research suggests that exposure to natural settings can have positive effects on mental health, including reduced stress levels, improved mood, and enhanced cognitive function. Individuals who have regular access to nature may engage in outdoor activities, exercise, and relaxation techniques that contribute to overall well-being.

In recent years, environmental inequality researchers have conducted numerous studies to determine whether poor people and racial and ethnic minorities are disproportionately burdened by residential proximity and exposure to environmental hazards (Bowen, 2002; Brulle & Pellow, 2005; Downey, 1998, 2003; Mohai & Bryant, 1992; Morello-Frosch, Pastor, & Sadd, 2001; Pellow & Brulle, 2005; Pellow & Park, 2002). Although this research has increased our understanding of race- and class-based environmental inequality in many important ways, it largely disregards the question of whether other subordinate groups, in particular children, are disproportionately burdened by, or vulnerable to, environmental inequities (Downey, 2005a; Shibley & Prosterman, 1998; for exceptions, see Bullard, 1983; Grineski, 2006; Pastor, Sadd, & Hipp, 2001; Pastor, Sadd, & Morello-Frosch, 2002; Stretesky & Lynch, 2002). Moreover, despite the dual emphasis in early environmental justice activism on subordinate groups' overexposure to pollution and underexposure to nature and green spaces (Downey, 2005b), few studies examine race or class inequities in access to nature and green space.

Although proximity to nature and exposure to green spaces may seem negligible compared with unequal pollution exposure and disparate proximity to toxic hazards, recent findings indicate that limited exposure to nature and green space may have serious physical and psychological health ramifications (Faber Taylor & Kuo, 2006; R. Kaplan & Kaplan, 1989; S. Kaplan, 1995; Kellert, 2005). Moreover, research suggests that children disproportionately suffer the long-term developmental consequences of limited experiences in nature (Kellert, 2005), making it imperative that environmental inequality researchers place children and their differential contact with and experiences in nature at the forefront of their research.

To make this case, this article highlights current research findings from the environmental health, environmental education, and environmental psychology literatures regarding the cognitive, emotional, and physical importance of childhood exposure to nature. This article also summarizes the few studies that have examined class- and race-based inequalities in children's exposure to the natural world and industrial environmental hazards. By synthesizing findings across multiple disciplines, we ultimately seek to provide a solid foundation for understanding the central environmental inequalities facing children and aim to convince environmental inequality researchers of the importance of investigating children's differential exposure to nature, green spaces, and industrial environmental hazards. Incorporating environmental health, environmental education, and environmental psychology research into the larger body of environmental inequality studies is important for several reasons.

It is widely accepted that children are more vulnerable than adults to the negative effects of environmental toxins because of the fundamental differences in children's physiology, metabolism, absorption, and exposure patterns that cause children's bodies to react to and excrete toxins differently than adults (Crom, 1994; Pastor *et al.*, 2002).

1. Increasing evidence suggests that access to nature and green space provides children with a myriad cognitive, emotional, and physical benefits, such as increased ability to concentrate, improved academic performance, reduced stress and aggression levels, and reduced risk of obesity
2. Unprecedented numbers of children suffer from asthma, cancer, low IQs, and learning disabilities each year (American Lung Association, 2007; Center for Health, Environment, and Justice, 2001; Louv, 2005). Although there is still much uncertainty as to the direct and indirect causes of these alarming health trends, current findings from the environmental psychology and environmental health fields are beginning to link children's exposure to environmental pollutants and diminishing access to nature to children's declining mental and physical health
3. Research has shown that regardless of race, ethnicity, and socioeconomic status, early childhood experiences in nature significantly influence the development of lifelong environmental attitudes and values (Chawla, 1998, 1999, 2006a, 2006b; Wells, 2000). As a result, many educators, environmentalists, scholars, and parents are becoming increasingly concerned about whether today's "de-natured" children will want to protect and care for the natural environment (EcoAmerica, 2006; Louv, 2007; Pyle, 2002; White, 2004).
4. Evidence suggests that the Govt is more concerned about children's health, safety, and general well-being than it is about other social concerns (Lucile Packard Foundation for Children's Health, 2007).

Finally, the negative impact of pollution exposure and limited access to nature on children's educational achievement, cognitive functioning, and mental health has serious implications for the future economic success of the United States in a highly competitive global market (Center for Health, Environment, and Justice, 2001). Therefore, focusing research efforts on children may be particularly useful in encouraging public policies and regulatory change dedicated to reducing pollution emissions, increasing access to nature and green spaces, and overcoming the pernicious effects of environmental inequality.

Urbanization and Built Environment: Urban environments, characterized by high population density, infrastructure, and cultural diversity, shape human behavior in various ways. The built environment, including architecture, transportation systems, and public spaces, can influence social interactions, mobility patterns, and lifestyle choices. Urbanization may also lead to changes in social norms, values, and community dynamics as individuals adapt to urban living conditions.

2.3 Behavioural Development-role of physical ecological factors, terrain, climate changes, demographics and their effect in human development.

After birth, there is striking biological and functional development of the brain's fiber tracts as well as remodeling of cortical and subcortical structures. Behavioral development in children involves a complex and dynamic set of genetically guided processes by which neural structures interact constantly with the environment. This is a protracted process, beginning in the third week of gestation and continuing into early adulthood.

Stages of Behavioral Development

Infancy (0-2 years): This stage is all about rapid growth and development, as our tiny humans learn to make sense of the world around them. From those first tentative smiles to the triumphant first steps, infants are constantly absorbing information and developing new skills. It's during this time that behavioral milestones such as object permanence and stranger anxiety begin to emerge, laying the groundwork for future cognitive and social development.

Early childhood (2-7 years): This stage is characterized by an explosion of language development and imagination. Children at this age are like little sponges, soaking up information and testing the boundaries of their world. It's also the time when they start to develop a sense of self and begin to understand the concept of rules and consequences.

Middle childhood (7-11 years): This is a time of increasing independence and social awareness, as kids learn to navigate friendships, academic challenges, and extracurricular activities. Cognitive abilities continue to develop, allowing for more complex problem-solving and abstract thinking.

Adolescence (11-18 years): It is a time of hormonal upheaval, identity crises, and questionable fashion choices. This stage is marked by significant physical, emotional, and social changes as young people transition into adulthood. It's a period of exploration and risk-taking, as teens push boundaries and seek to establish their own identities separate from their parents.

Adulthood (18+ years): Brains continue to develop well into our twenties, and we continue to learn and grow throughout our lives. Adulthood brings new challenges and opportunities for growth, from career development and romantic relationships to parenthood and beyond.

Theoretical perspective

Behaviorism (the school of developmental psychology based on behavioral models) offers alternative ways of how development process take place. There are several different behavioral models not just one. All behaviorists share the same basic tenet. Development is a process that adheres to the **laws and principles of learning. Biological elements** provide some of foundation for the development but **experiences and opportunities** to learn are the central and necessary elements that give direction to the process. Behaviourism began in the late nineteenth century with the work of Russian psychologist Ivan Pavlov (1849-1936), and was further developed in the United States by Edward L. Thorndike (1874-1949), John B. Watson (1878-1958) and B. F. Skinner (1904-1990). One of its most famous principles is the idea that all behaviour, in both animals and humans, can be traced back to a neurological association between stimulus and response.

Piaget's Cognitive Development Theory: Jean Piaget, the Swiss psychologist with a penchant for watching children play, gave us a framework for understanding how our thinking processes evolve over time. His theory suggests that children actively construct their understanding of the world through four distinct stages: sensorimotor, preoperational, concrete operational and formal operational. It's like watching a little scientist at work, experimenting and forming hypotheses about how the world works.

Erikson's Psychosocial Development Theory: Erik Erikson took a more holistic approach, focusing on how social relationships shape our development throughout our entire lives. His theory outlines eight stages, each characterized by a specific psychosocial crisis that must be resolved. It's like a video game, where each level presents a new challenge to overcome before moving on to the next.

Bandura's Social Learning Theory: Albert Bandura brought us the idea that we learn not just through direct experience, but also by observing others. This theory emphasizes the importance of modeling and imitation in behavioural learning. It's the reason why kids might suddenly start using their parents' favorite phrases or why we might pick up new habits from our friends.

Bronfenbrenner's Ecological Systems Theory: Urie Bronfenbrenner took a step back and looked at the bigger picture, proposing that development occurs within a complex system of relationships and environments. His theory considers the various layers of influence, from immediate family to broader cultural contexts, that shape our development. It's like a set of Russian nesting dolls, with each layer representing a different level of environmental influence.

Vygotsky's Sociocultural Theory: Lev Vygotsky emphasized the role of culture and social interactions in cognitive development. He introduced the concept of the "zone of proximal development," which describes the sweet spot between what a child can do independently and what they can achieve with guidance. It's like having a personal trainer for your brain, pushing you just beyond your comfort zone to foster growth.

Basic assumptions of behavioral Models

These theories share some basic tenets that serve as starting points for most developmental behaviorists. This foundation shared by behaviorists provides some continuity between past and present behavioral prospective. In fact the links between today's behavioral theorists and the behaviorists of the early 20th century are stronger than similar links found in other theoretical perspectives.

All the behavioral theories have the basic assumptions in common.

1. **Development is a function of learning**: Robert Gagne defines behavioral development as "**the cumulative effects of learning**". Learning is defined as short term changes in behavior. When these short term changes are combined and are hierarchically organized i.e. they build on each other, the result is development. Development then is the result of the linking of accumulated experiences. **Development results from learning- learning does not result from development.**
2. **Development is the result of different types of learning**- Each type of learning governs the developmental process. eg., classical conditioning, Stimulus-Response connection, operant conditioning, etc.
3. **Individual differences in development reflect differences in the histories and past experiences of individuals**- Differences in individual development result from the past experiences that people have had and the value of these experiences for helping the individuals to accomplish immediate or long term goals. The history of the individual according to behaviorists is series of short term changes that when linked together form the foundation for development. The content of these experiences and the way they are organized contribute to the differences between one person's developments from the other person. Eg:- child rearing is different for different people. These experiences are part of the individual's history of learning. The dynamics of how people store and recall their experiences and how they transfer their experiences to new situations are crucial elements of the behavioral perspective.
4. **Development results from the organization of existing behaviors**- Simple isolated behaviors are organized in to complex patterns. Reflexes nearly 30 of them is the individual elements that form the basis for more complex behaviors. Later development with all its richness and complexity is related to these early biological reflexes. Over time the different experiences of the infant and children result in the unique

organization of the biological potentials of each child. As they grow the differences among individuals become even greater because the organization of their experiences undergoes further change.

5. **Biological factors set general limits on the kinds of behaviors that develop**, but the environment determines the behaviors in which the individual engages. Although biological processes provide the framework, the environment determines the kinds of behavior an individual will acquire. Even conception is not purely a biological event, because environmental factors like hospitality of the womb, maternal diet, smoking, alcohol intake, etc can effect development significantly. From a behavioral perspective, if biological needs (such as good health care, nutrition and exercise) are met, development progress depends on sequencing and relevance of experiences.
6. **The development of the individual is not directly related to biologically determined stages**. Behavior is not biologically determined, nor is it necessarily tied to or a function of internally regulated biological processes. Behaviorists, however, do not dismiss obvious universal maturational sequences. Being able to maintain balance, for example, is a pre-requisite of walking. However, if a child who is capable of maintaining balance is not given the opportunity to learn to walk, walking may be significantly delayed. On the other hand, the provision of certain types of experiences can accelerate development.

The six assumptions discussed above are not universally held among behaviorists. Because behaviorism is an evolving point of view, some advocates of behaviorist models would qualify these assumptions in various ways. Although they are not hard-and-fast rules, they illustrate the behaviorist viewpoint in general and can be used as a tool for organization.

Factors Influencing Behavioral Development: Nature vs. Nurture

The age-old debate of **nature versus nurture** has evolved into a more nuanced understanding that both genetic predispositions and environmental influences play crucial roles in shaping our behavior. Let's explore some of the key factors that contribute to the complex tapestry of human development.

Genetic predispositions: Our genes provide the blueprint for our development, influencing everything from our physical characteristics to our temperament. Some traits, like eye color, are directly determined by our genes. Others, like personality traits, may be influenced by genetic factors but can be modified by environmental experiences. It's like starting a game with a particular character class – you have certain inherent strengths and weaknesses, but how you develop those traits is up to you.

Environmental influences: The world around us plays a huge role in shaping our behavior and development. From the physical environment (like access to **nutrition and healthcare**) to the social environment (like **exposure to language and cultural practices**), our surroundings constantly mold our experiences and, consequently, our development.

Parenting styles and family dynamics: The way we're raised has a profound impact on our development. Authoritative, authoritarian, permissive, or neglectful parenting styles can lead to different outcomes in children's behavior and emotional well-being. Family dynamics, including sibling relationships and extended family interactions, also play a crucial role in shaping our social skills and emotional intelligence.

Peer relationships and social interactions: As we grow, our social circle expands beyond the family unit. Friendships, romantic relationships, and interactions with classmates and colleagues all contribute to our ongoing behavioral development. These relationships provide opportunities for social learning, emotional growth, and the development of important life skills.

Cultural and societal norms: The broader cultural context in which we live shapes our values, beliefs, and behaviors in profound ways. From gender roles to religious practices, cultural norms influence how we interpret the world around us and how we interact with others. It's like the operating system of our social world, running in the background and influencing everything we do.

Milestones in Behavioral Development: The Building Blocks of Growth

Language acquisition: From those first adorable babbles to complex sentences, language development is a fascinating process. It's not just about learning words – it's about understanding the rules of grammar, the nuances of communication, and the social aspects of language use. The ability to express ourselves and understand others is fundamental to our social and cognitive development.

Emotional regulation: Learning to manage our emotions is a crucial skill that develops throughout childhood and continues into adulthood. It involves recognizing and labeling emotions, developing coping strategies, and learning to express feelings in appropriate ways. This skill is essential for maintaining healthy relationships and navigating the ups and downs of life.

Social skills development: From learning to share toys in preschool to navigating office politics as an adult, social skills are constantly evolving throughout our lives. These skills include empathy, cooperation, conflict resolution, and the ability to read social cues.

Moral reasoning and ethical behavior: As we grow, we develop an increasingly complex understanding of right and wrong. This process involves learning societal rules, developing a personal moral code, and grappling with ethical dilemmas. It's like upgrading our internal compass, allowing us to navigate increasingly complex social situations.

Self-concept and identity formation: Who am I? This existential question is at the heart of identity formation, a process that begins in childhood and continues throughout our lives. It involves developing a sense of self, exploring different roles and identities, and integrating various aspects of our personality into a coherent whole. It's like creating a character in a role-playing game, but with real-life consequences!

Implications of Behavioral Development: Why It Matters

Educational strategies and interventions: Insights from behavioral development research can inform educational practices, helping teachers tailor their approaches to meet the needs of students at different developmental stages. For example, understanding the importance of play in early childhood development has led to more play-based learning approaches in preschools and kindergartens.

Mental health and well-being: Knowledge of typical developmental patterns can help identify potential issues early on, allowing for timely interventions. Understanding the foundations in behavior can also inform therapeutic approaches and strategies for promoting mental health across the lifespan.

Career and relationship success: Awareness of our own developmental journey can help us make more informed choices about our careers and relationships. Understanding how our past experiences shape our current behavior can lead to better self-awareness and improved interpersonal skills.

Parenting and child-rearing practices: Armed with knowledge about behavioral development, parents can create environments that support their children's growth and learning. This might involve adapting parenting styles to meet the changing needs of children as they grow, or understanding the importance of certain types of play or social interactions at different ages.

Social policy and community support: On a broader scale, understanding behavioral development can inform social policies and community programs. For example, recognizing the importance of early childhood experiences has led to increased investment in early education programs in many countries.

Terrain: Effect of physical topography on human behavior

According to a recent body of work, environmental and climatic changes played an important role in early human migration, choice of habitats , species extinctions, interbreeding , and natural selection. To make such inferences and identify the external drivers of human evolution on a large scale, it is necessary to link archeological and anthropological datasets to reconstructions of regional environments and climates. Previous studies focused on the influence of climatic and ecological variables on the development of human population structure , for example, temperature, precipitation, net primary production, and biome types. All these climatic factors are further affected by landscape features, such as elevation and topographic roughness.

In particular, mountain slopes are places where microclimates and multiple ecotones can coexist in close proximity, creating biodiversity hotspots for flora and fauna . Such habitats could have provided a wide range of food resources for well-adapted early hunter-gatherer populations. Recent studies demonstrated that terrain slope emerged as an important predictive parameter for constructing ecological niche models for Homo sapiens during the Last Glacial Maximum and Marine Isotope Stage 3, as well as for Homo neanderthalensis(Neanderthal, (Homo neanderthalensis, Homo sapiens neanderthalensis), member of a group of archaic humans who emerged at least 200,000 years ago during the Pleistocene Epoch (about 2.6 million to 11,700 years ago) and were replaced or assimilated by early modern human populations (Homo sapiens) between 35,000 and perhaps 24,000 years ago.) during the Last Interglacial. In addition, other studies focusing on Africa have pointed out that there is a high number of early human fossils and artifacts located in areas with increased roughness due to access to water and the distribution of vegetation.

The physical geography of an area plays a crucial role in determining the resources available and the constraints that define human activity.Physical features such as vegetation, climate, water cycle, and land formations not only determine whether humans can live in a certain area but also influence their lifestyles. People adapt to the available food sources and climate patterns, shaping their daily routines and cultural practices. Physical features directly affect the distribution of human settlements. Factors such as altitude, latitude, land forms, climate, and soil condition determine where permanent settlements can be established. High altitude, for instance, limits large permanent settlements due to the lack of oxygen. Physical systems, such as the earth's processes and features, create, maintain, and modify the environment in which humans live. Understanding these processes can greatly influence people's decisions about where they choose to reside, work, travel, and even what they eat.Terrain features such as mountains, rivers, forests, and deserts play a significant role in shaping human behavior. Geographic barriers can affect patterns of settlement, migration, and trade, leading to the formation of distinct cultural groups and social networks. For example, mountainous regions may foster isolation and cultural homogeneity, while coastal areas may encourage trade, exploration, and cultural exchange.

Regional differences in personality are associated with a range of consequential outcomes. Frontier settlement theory suggests that physical topography is a crucial factor shaping the psychological landscape of regions. Gotx *et al*, (2020) investigated whether topography is associated with regional variation in personality across the United States (n=3,387,014). Consistent with frontier settlement theory, it was revealed that mountainous areas were lower on agreeableness, extraversion, neuroticism and conscientiousness but higher on openness to experience, mountainousness as a meaningful predictor of personality when tested against a conservative set of controls. East–west comparisons highlighted potential differences between ecological (driven by physical features) and sociocultural (driven by social norms) effects of mountainous terrain.

In the United States, Oishi and Tsang (2022) found that low-mobility areas see increased trust in familiar friends but limited interactions with strangers, resulting in lower overall trust levels. Conversely, regions with medium mobility in the United States exhibit higher trust levels, indicating a positive link between residential mobility and generalized trust. In contrast, Yuan, Yang, Sun, Yin, and Li (2021) reported a negative correlation in China, showing that higher residential mobility corresponds to lower generalized trust levels. They suggested that frequent changes in residence contribute to heightened perceptions of uncertainty, which in turn reduce generalized trust in China.

Psychological outcomes are associated with different environmental factors, including physical geography (Stieger *et al*., 2022), climate (Wei *et al*., 2017), terrain (Kitayama, Ishii, Imada, Takemura, & Ramaswamy, 2006), and land usage (Militaru *et al*., 2023). Among them, mountainousness (associated with rugged terrain, steep slopes, and high elevations) is an important topographical factor affecting different psychological traits (Götz *et al*., 2020; Stieger *et al*., 2022; Xu *et al.,* 2022). **First**, the demanding conditions prevalent in mountainous areas could exert an impact on generalized trust. Some research suggested that mountainous areas are often harsh and inhospitable (Kitayama *et al.*, 2006, 2010), which leads to a shortage of survival resources. Limited resources may make residents distrusting strangers and being less likely to behave prosocially (Sng, Neuberg, Varnum, & Kenrick, 2018). On the other hand, other researchers have suggested that the challenging and isolated nature of living in mountainous regions may lead to higher levels of cooperation and interdependence among community members (Fuchs, Röthlisberger, Thaler,

Zischg, & Keiler, 2017; Xu *et al.*, 2006), which could foster a greater sense of trust.

Second, mountainous regions are intertwined with social-level factors like GDP and residential mobility, which significantly shape generalized trust. China exhibits a distinct regional contrast, with the eastern part experiencing stronger economic growth compared to the western counterpart (Zhou *et al.*, 2022). Zhu, Fang, Yang, Qiu, and Yu (2018) underscored altitude's impact on China's economic development, revealing lower income levels concentrated in the western high-altitude mountain areas. The limited GDP in these regions could contribute to lower generalized trust among residents, as prior studies found a positive link between China's economic performance and trust (Niazi & Hassan, 2016). Additionally, the rugged terrain in mountainous areas can lead to inconvenient transportation, restricting people's mobility between areas. Reduced resident movement could lower perceived uncertainty, potentially fostering increased generalized trust among the Chinese population (Yuan *et al.*, 2021).

Third, prior research has demonstrated that China's mountainous regions exhibit elevated levels of agreeableness and openness to experience (Xu *et al.*, 2022), traits that have been strongly linked to trust in strangers (Freitag & Bauer, 2016). Consequently, it is plausible that China's mountainous areas might also manifest heightened levels of generalized trust. Since the relationship between mountainousness and generalized trust is complex and not yet fully understood, more research is needed to determine the nature and extent of this association.

Climate Changes: In the late 19th century, the scientist (and suffragette) Eunice Newton Foote published a paper suggesting that a build-up of carbon dioxide in the Earth's atmosphere could cause increased surface temperatures. In the mid-20th century, the British engineer Guy Callendar was the first to concretize the link between carbon dioxide levels and global warming. Now, a century and a half after Foote's work, there is overwhelming scientific evidence that human behavior is the main driver of climatic changes and global warming. The negative effects of rising temperatures on the environment, biodiversity and human health are becoming increasingly noticeable. The years 2020 and 2016 were among the hottest since the record keeping of annual surface temperatures began in 1880. Throughout 2022, the globe was plagued by record-breaking heatwaves. Even regions with a naturally warm climate, such as Pakistan or India, experienced some of their hottest days much earlier in the year — very probably a consequence of climate change. According to the National Centers for Environmental Information of the United States, the surface global temperature during the decade leading up to 2020 was +0.82 °C (+1.48 °F) above the 20th-century average. It is clear that we are facing a global crisis that requires urgent action. Human behaviour is not only one of the primary drivers of climate change but also is equally crucial for mitigating the impact of the Anthropocene (the current geological age, viewed as the period during which human activity has been the dominant influence on climate and the environment). The complexity of humans is also reflected in their psychology. The Historical Perspective: What Can We Learn from the Past? Has been presented in the give table (1).

Psychologists are not only concerned with understanding and addressing climate scepticism but are also increasingly worried about mental health consequences for which individual- and system-level behaviour change is necessary.

Extreme climate events (ECEs): They are the consequences induced by climate change (e.g., floods, droughts, heat waves, storms). Unlike natural disasters, they are not occasional events in a stable system, whereas they are more frequent and extreme and long-lasting. However, there is still no consensus on the definition, since it is not clear how to classify their extremes, frequency, and impacts.

2. Vulnerability and Resilience

Vulnerability to climate change is defined as "the degree to which a system is susceptible to, or unable to cope with, adverse effects of climate change". Health risks associated with climate change vary depending on multiple factors, such as the nature of the exposure (e.g., the location of a population exposed to drought or flooding), the associated hazards (e.g., projected change in precipitation patterns or climate suitability for infectious disease transmission), socio-economic and environmental determinants for the population and individual (e.g., age, gender, water coverage, sanitation, and hygiene systems), and the capacity of health systems to protect against current and future risks. Indeed, some populations are disproportionately disadvantaged in dealing with climate change and may more easily develop **stress-related psychopathological reactions** and/or disorders due to their geographical position (coasts, areas with intense hurricane activity or subjected to heat waves), to involvement in climate-sensitive activities (agriculture, aquaculture, fishing, pasture) or to reliance on natural resources and ecosystem services for their lives.

Species interactions, emergent properties, and biological invasions

Variability in species' exposure and responses to climate change are primary drivers of altered species interactions. Emergent properties of ecosystems, including community characteristics such as food-web structure and function that are mediated by species interactions, are changing as species shift their distributions. Many non-native invasive species are opportunistic generalists that can take advantage of changing conditions, colonize disturbed areas, and out-compete species, thereby altering community composition, dominance, production, and increasing extinction risk in some cases.

On the other hand, farmers successfully coping with climate change consequences and integrating into a larger economy may nonetheless be affected by market failures caused by unfair and unpredictable patterns of globalized trade and become a vulnerable group. In general, poverty is an important condition linked to vulnerability to climate change, as it is directly associated with limited resource access, and women, especially in poorer countries such as India, China, and Brazil, are the more affected. This can lead to a **feminization of poverty** and an increasing trend of **male out-migration** for remittances.

Children and the elderly must be considered too: Children show vulnerability especially in terms of long-lasting or irreversible outcomes, showing post-traumatic reactive phenomena; during disasters, they can be separated from their families, schools, or childcare centres, and at a later time, it can be harder for them to continue their education or they may show poorer academic performance, while elderly people are notoriously attached to their environment and share the same vulnerabilities of individuals who are disabled, chronically ill, or with pre-existing medical conditions. Other vulnerable groups include people with mental illness, homeless, minorities, migrants, and refugees that generally have limited access to resources and lack the adaptive capacity to protect themselves, with the risk of negative mental health outcomes. Moreover, many environmental events can cause displacement, but the concept of a climate displaced person is still vague with no clear legal protection, so **that climate migrants might struggle with isolation, unreliable living and working conditions, or discrimination, as well as frequent physical and mental health problems**. Particular attention should be given to indigenous communities because they are exposed to more than one factor of vulnerability and offer a striking example of coping difficulties. They are often located in geographically peripheral and vulnerable regions, several communities face ethnic discrimination, racism, prejudice, bullying, disempowerment, and grief associated with the loss of homelands and their traditional way of life. Vulnerability stems from colonial intrusion into a traditional lifestyle, rapid modernization, and cultural assimilation policies, which cause a loss of traditional knowledge and limit traditional adaptation strategies. All these circumstances may be worsened by climate change affecting lifestyle and health, both physical (altering outdoor activities and diet) and mental (inducing stress, worry, depression, anxiety, trauma, and cognitive biases), but also resulting in spiritual loss caused by the transformation or the disappearance of sacred ritual important sites and the disrupted relationship with the land that is the source of ancestral linkages, even leading to high rates of substance abuse, violence, and suicide.

On the other hand, **resilience** is defined as the "**capacity of a social-ecological system to cope with a hazardous event or disturbance, responding or reorganizing in ways that maintain its essential function, identity, and structure, while also maintaining the capacity for adaptation, learning and transformation**", the maximum perturbation a system can tolerate before a transition to an alternative stable state. At an individual level, resilience depends on inherited biological and psychological factors involving personal history, skills, experiences, cognitive and behavioural efforts, social context, and occurrence of adverse events and allows us to cope with adverse events, possibly resist changes, and recover from distress. It has been argued that resilient people anticipate risks, reduce vulnerability to those risks, respond effectively to threats, and recover faster, thus increasing the capacity to respond to subsequent threats. On a social level, resilience is the ability of communities to prevent, recognize, and confront uncertainty, it enables adaptive preparation for future events, responds to damaging events, mitigates external shocks, deals and manages the changes to infrastructure and the environment, provides aid and services to disaster victims, directs economic and social systems to enable recovery.

3. Adaptation

Natural disasters are known to be accidental and to follow a predictable course after their manifestation; as a result, they are typically associated with better recovery. In contrast, human-induced or technological disasters are typically associated with a more complex and painful recovery, especially when those disasters are perceived to be preventable and avoidable. **Adaptation is about processing information and how to practically prepare for extreme events and to a changing environment**. Adaptation to climate change requires from humans a frequent switch between different levels of perspective: **macro** (DNA, resources, possibilities of species), **meso** (collectivity, ethnic realities, mass and psychosocial phenomena), and **micro** (psychological adaptation, family and individual). There are **three ways of adapting to climate change:genetics, behaviour, and migration**, depending on intrinsic characteristics, such as genetic architecture, behavioural flexibility, life history, demographic factors, and dispersal capacities. However, genome and migration can be considered a single endogenously interrelated system, since successful adaptation has been associated with genes involved in exploration and dispersal capacity. Genetic evolutionary properties and/or behavioural adaptation factors allow species to resist in their current geographic range, while dispersal potential and migration ability allow adaptation in different geographic conditions, since it improves the access to resources. For humans, this phenomenon is called "**ecomigration**" and can also result in displacement and relocation, leading to "**environmental refugees**".

4. Emergent Behaviours, Emotional Regulation, and Mass Effect Phenomena.

In an adaptation effort, human collective behaviours may emerge when events are perceived as overwhelming or when the control in the use of information technology is lacking. These behaviours have also been called "emergent behaviours", because they are not previously known or observed, and they arise in complex systems from the interaction of different parts, none of which displayed that behaviour individually. They can be either beneficial or potentially harmful. Climate change information has different influences on individuals' psychological believes, and the kind of climate change events is linked to different emotional regulation styles. Sometimes, people may be distressed about climate change without knowing it, sensing only a vague unease about the changes happening around them. The direct experience of repeated extreme climatic events on one's territory can trigger a stronger reaction because the threat is perceived as concrete and close, whereas climate change is more abstract and distant if it is not directly witnessed; therefore, a reduced "**psychological distance**" is associated with a higher level of concern and more pro environmental intentions that lead to better adaptation.

5.The Climate Change–Violence Model: The Climate Change–Violence Model (see Figure below ; Miles Novelo & Anderson, 2019) highlights how rapid climate change would influence human aggression and violence. The model demonstrates that there are three major pathways through which rapid climate change will increase human aggression and violence through one direct and two indirect mechanisms. The direct mechanism is known as "the heat effect." This is the wellreplicated finding that as people become "uncomfortably" hot, they become more irritable, perceive other people's behavior as more threatening and aggressive, think more aggressively, and behave more violently (Anderson, 2001).

The indirect mechanisms (from things such as failed crops, water shortages, mass migration, and political instability) stem from (a) developmental factors that will increase the likelihood of children becoming violence-prone adolescents and adults (e.g., malnourishment), and (b) factors that increase intergroup conflicts (e.g., mass migration).

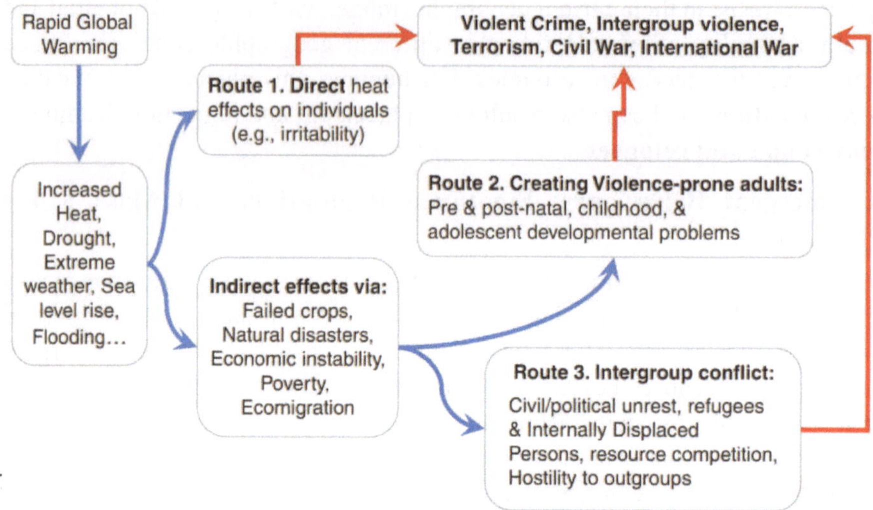

5. Mental Maladjustment and Socioeconomic Processes

The World Health Organization has defined **climate change as one of the greatest health threats** of the 21st century and an increase in climate-change-related events will entail higher risks to human health and survival, also exacerbating the already increasing incidence of **non-communicable diseases** NCDs (i.e., chronic diseases of long duration and the result of a combination of genetic, physiological, environmental, and behavioural factors) that include **mental disorders**. The impact of climate change on people's physical, mental, and community health can arise both directly and indirectly. The direct effects on mental health may occur rapidly, usually from extreme weather events and natural disasters, or gradually as slowly progressive but not necessarily life-threatening (e.g., changing temperature and rising sea levels). The indirect effects can be caused by poor physical health (which is associated with mental wellbeing), by environmental risk factors (such as smoke, pollen density, dust, plant disease, infestations, availability of water, water disease, living in urban slums, loss of sense of place), through the social environment (their impact on human activities), via adaptation and mitigation (e.g., travel by alternative means, availability of air conditioning). Moreover, climate change may have an impact on individuals and communities not directly affected by it, including emotional and affective responses, which can arise when observing climate change effects worldwide, viewing images of environmental degradation or human suffering in the media, questions of lifestyle or purchasing choices, usually being influenced by personal values, beliefs, and experience, and by social representations of climate change and its impacts.

Mental health effects of climate change range from **minimal stress to clinical disorders**, and they can be classified as **acute or chronic, bearing in mind that some consequences may overlap**. Acute effects are usually consequences of extreme and powerful weather events or natural disasters (such as wildfires, tornadoes, hurricanes, storms, floods, and droughts) that may occur with little or no warning, involving loss of life, resources, social support, and social network. Generally, such reactions can be diagnosed as acute stress disorders if they begin within four weeks after a disaster or, if longer, as post-traumatic stress disorders (PTSD), depression, and anxiety. Many exposed people exhibit hostility, paranoia(unreasonably or obsessively anxious, suspicious, or mistrustful), physiological hyperarousal, chronic dissociation, detachment, cognitive symptoms, poor quality of sleep, increased domestic violence, alcohol, substance abuse, psychosomatic disorders, and suicidal thoughts.

Oppressive heat waves may induce exhaustion, increased suicide rates, higher rates of self-harm, increased mortality and morbidity risks amongst psychiatric patients, as well an increased utilization of mental health services. Chronic effects result from long-term changes in climate. In addition to PTSD, depression, and anxiety, it is possible to detect the presence of post-disaster adjustment, loss of identity, loss of autonomy and control, and many emotions such as fear, anger, helplessness, resignation, exhaustion. In addition, there are large-scale social and community effects, even taking the form of violence, conflicts over scarce resources, displacement, and migration. Effects can be delayed and may also persist over several years affecting communities in broader areas as well.

Children can also be affected, showing posttraumatic reactive phenomena, such as anxiety disorders and panic attacks, sleep problems, adjustment disorder acute stress reactions, compulsively repetitive play, re-experiencing fears, and psychotic disorders. In some papers, the association between climatic events and mental disorders has been labelled with new terms as eco-anxiety, eco-guilty, ecoparalysis, ecological grief, and solastalgia. Psychotherapists are pioneering a new field of treatment termed "**ecopsychology**". "**Ecoanxiety**" can be described as anxiety related to a changing and uncertain environment and becoming overwhelmed by the sheer scale and complexity of the problem faced. "Ecoparalysis" refers to apathy and disengagement with reality due to the inability to meaningfully respond to the climatic and ecological challenges, in contrast to expressions of avoidance. "Ecological grief" is the grief associated with physical ecological losses (land, ecosystem, and species) and attendant ways of life and culture, or the grief associated with disruptions to environmental knowledge systems and resulting feelings of loss of identity, or the grief associated with anticipated future losses of place, land, species, and culture, due to acute or chronic environmental changes. The term "**solastalgia**" refers to the pain or sickness caused by the loss or lack of solace from one's home and territory, with the awareness that the living and loved place is in danger, leading to the fading of the sense of belonging (identity) to that particular place and a feeling of distress (psychological desolation) about its transformation. In contrast to "**somaterratic illnesses**" (soma = body, terratic = Earth-related) that threaten physical wellbeing and are caused mainly by living in ecosystems that have been contaminated by pollutants and toxins, these new concepts can be synthesized as a form of "**psychoerratic illnesses**", defined as an Earth-related mental illness where people's mental wellbeing (psyche) is threatened by the severing of "healthy" links between themselves and their home/territory.

Leveraging technology and infrastructure

Human communities may need to adapt by updating technology and infrastructure. Extreme events often motivate adaptation and Strengthening governance. Promoting local climate adaptation initiatives can also be an effective management strategy. Self-determination and self-governance of tribes and Indigenous peoples, as well as involvement of communities of color or natural resource users such as anglers in particular localities, supports the use of local knowledge and cultural practices that can track, adapt to and mitigate environmental change.

Urban ecosystems provide residents with many benefits, or "ecosystem services" Examples include urban forests reducing air temperatures, urban streams and floodplains regulating stormwater , gardens and urban farms producing food, and parks providing space for recreation and exercise . As the proportion of the global population living in urban areas continues to grow, the importance of urban ecosystem services for human well-being will increase. Cities around the world vary widely in terms of climate , socio-economic conditions , and demography , and this variation influences the types of ecosystem services that are most needed. Cities are located in most of the world's climate zones , and this wide variation in climate has implications for which ecosystem services are important in any particular region. The cooling effect of vegetation, for example, will be most valued in cities where high temperatures and humidity reduce thermal comfort or increase the risk of heat-related mortality. Cities also vary greatly in their economic status, from the wealthiest financial and industrial centres to the poorest developing economies. This gradient in prosperity—which is closely linked to human development—also influences the relative importance of ecosystem services. For example, urban and peri-urban agriculture is of minor significance as a source of food in the developed world but is essential to many residents in poor cities. Finally, urban areas vary in population density, from dispersed low-rise settlements to compact, high-rise districts.

Implications: Reactions to extreme weather events are similar to traumas from natural disasters, while long-term changes are more challenging. The pressure of climate change will possibly lead to new forms of mental distress, climate-related mental disorders (CRMD), and to new adaptations. Nonetheless, if changes in climate will keep an instable course with more frequent extreme events, we can assume that repeated failed efforts to adapt (at least for a while) will lead to a stratification of chronic distress and unsuccessful coping strategies.

2.4 Children's physical environment, housing, overcrowding, neighborhood and development. Economic factors and their impact on human development

Housing identifies five significant domains of health child development: physical health and well-being, social knowledge and competence, emotional development and maturity, language and cognitive development, healthy physical development, and communication skills and general knowledge. The interdependency of many of these domains is emphasized, and one important underlying determinant of these domains common to families in lower socioeconomic strata is the toxic stress of poverty . At the same time, poverty is an important determinant of poor-quality and unhealthful housing; indeed, housing may be one of the important ways that poverty shapes experiences that lead to toxic stress for children, both directly and indirectly (through parental and family interactions).

Factors moderating the relationship between housing and healthy child development

- Household and demographic characteristics include the following
- Housing tenure (own/rent)
- Income level
- (Dis)ability
- Age

- Gender
- Race/ethnicity
- Household type

Housing can have a potentially direct effect on child health in several ways, through biological, physical, and chemical mechanisms . A large body of research draws quite direct relationships between exposure to biological (e.g., mold, infestations, vermin), physical (e.g., disrepair, hazards, temperature extremes), and chemical (e.g., radon, chemical exposures) hazards on the one hand and elevated risk of physical illness on the other hand. Structural deficiencies in housing are responsible for a large burden of morbidity in children, and the population exposed is very large and differentially distributed to the disadvantage of low-income, racialized, and marginalized households.

Poor housing structure leads to dust mite attack to children leading to allergies, poor bedding leading to asthma, dampness in housing leading to asthma and fungal infections, Poor physical design of housing may lead to elevated risk of trips, slips, falls, etc. and physical injury—a potential health hazard that features centrally in building codes, but such codes cannot protect occupants of older dwellings and dwellings that are in disrepair or have not been upgraded to current standards.

Studies have shown that having a place where children can go to be alone, despite crowded housing, can be psychologically protective because such a space allows household occupants to regulate social interaction and exposure to other stimuli in the housing environment (e.g., noise, light) . Interior design and layout of a dwelling, therefore, may affect health through the role it plays in providing an environment where occupants can practice self-regulation—a property of housing that is sometimes described as "restorative". Where this restorative quality is lacking, occupants experience additional stress and elevated risk of negative health impacts.

Psychological Attributes of Housing

Two key aspects of the psychological attributes of housing include meaning and control of housing.

Considering the importance of meaning first, the home is one of the most important sites for the investment of meaning in our lives, and a lack of meaning in life is a significant psychological risk for adults and children alike . Closely related to meaning is the concept of identity, which is a significant part of the relationship people have with their homes, as the home is an expression of self, an indicator of status, and a reflection of personal values . In terms of control, the social epidemiological literature suggests that having control over one's life is a significant determinant of health. Indeed, many have hypothesized that control is the underlying factor that explains the socioeconomic gradient in health, at least with adults and in worksites. The greater one's income, educational attainment, or other measure socioeconomic position, the greater control one is able to exercise over everyday life, and it is this underlying control that causes differentiation in health status by socioeconomic status.

Social Attributes of Housing

Social support is a significant determinant of health, and social isolation is a significant risk factor for poor health. Social support performs an important range of functions for adults, including guidance (advice or information), reliable alliance (assurance that others can be counted on in times of stress), reassurance of worth (recognition of one's competence), attachment (emotional closeness), social integration (a sense of belonging to a group of friends), and opportunity for nurturance (providing assistance to others). The existence of social support among adults is of course important for the children in their care in a domestic environment. The home is an important site for the establishment and maintenance of social ties.

Financial Attributes of Housing

Urban land markets act as important engines of economic inequality in market-based societies, exacerbating underlying inequalities that stem from the operation of the labor market. markets tend to transfer income in a regressive fashion, that is, from lower-income households to upper-income households. This transfer happens through a number of mechanisms, and most of these hinge on the distinction between homeowners and households who rent their housing. The tax advantages of home ownership are significant and largely invisible. The foregone government revenues from providing (largely tax) subsidies for homeownership are many times greater than spending on affordable housing in many countries. For this and other reasons, renter households, who tend to be lower income, tend to pay a greater proportion of their income toward shelter costs and have a limited number of affordable options available to them. Unaffordable housing is a significant threat to health. Studies find evidence of a relationship between **housing affordability and child cognitive achievement** and evidence of discounting of child well-being in the face of high housing costs as a proportion of income.

Locational Dimensions of Housing Land markets tend to operate as powerful socio-spatial sorting mechanisms, sorting people of similar socio-economic circumstances into similar parts of the city. This is significantly aided by zoning and other urban planning mechanisms, which govern proximity to amenities, size of residential lots. According to Ellen & Glied, low-income children consistently live in poorer neighborhood conditions than do other children, and these neighborhoods are characterized by a number of attributes that make them less healthful for children. First, evidence shows that low-income children are more likely to be exposed to pollutants in their neighborhood of residence and to experience negative health effects as a result. High levels of automobile traffic are also more common in low-income housing neighborhoods, leading to elevated air pollution and risks to pedestrians. A number of amenities that serve everyday life are either absent, too widely dispersed, more expensive, or of poorer quality in low-income neighborhoods, including grocery stores with affordable nutritious foods, recreational facilities, schools, and health care services. Systematic neighborhood disadvantages in health opportunity structures such as these are driven in part by racial and income segregation.

Housing crowding effects on children's wellbeing

One's housing relates to many aspects of social life including privacy, location, health, security, social relations, and community resources. One aspect of housing quality is the quantity of housing that is available to each member of a household. For a given household size, the size of the dwelling unit determines the degree of housing crowding experienced by the persons who live there. Living in crowded housing conditions can create stress in the home and have negative consequences for its inhabitants. Children may be particularly vulnerable to this type of poor housing quality because they use space in the home to do homework, interact with family members, develop an identity, practice skills, and sleep. Because the environments in which children are raised vary substantially across socioeconomic groups, these environments may also contribute to the intergenerational transmission of social inequality. Variation in housing crowding may be a key aspect of children's physical environments. This paper is based on an analysis of representative data for the US and for a large metropolitan area with atypically high levels of housing crowding on the academic achievement, behavior problems, and physical health of children.

Most research on the effects of crowding is based on adults, whose characteristics, including health, socioeconomic status, educational attainment, labor force participation, and homeownership, are a function of both current and past experiences in the home, workplace, and elsewhere (Evans, 2003; Newman, 2008). Yet children, on whom little crowding research has been done, are particularly dependent on and influenced by their home environment. The home is where the majority of children's socialization, skill development, and identify formation occurs. These processes can be disrupted if the home environment is strained by overcrowding.

Concerns regarding the consequences of population overcrowding began with Malthus in the late 1700s. It was in the early 1960s, however, when a flurry of empirical research on the negative effects of population overcrowding was sparked by findings from experiments on densely populated animals. Calhoun's study of laboratory rats found correlations between population density and increased aggression, disruptions in mating patterns and maternal activity, and higher rates of illness (Calhoun, 1962). Animal studies inspired investigations on the effects of crowding on humans; the results of these studies, however, were far less consistent than in those with animals.

The neighborhood is a child's proximate physical and social environment to which he or she is continually exposed. Although neighborhoods can serve as important environments for children, their effects are difficult to distinguish from those of their families whose resources and preferences determine where they live. Families affect children more strongly than neighbors because they are physically and socially closer and more homogeneous. Individuals are socialized within the home to learn roles and relate to others. Individuals eat, sleep, work, and seek privacy in the home. As a result, the physical characteristics of dwelling units generally, and their degree of crowding in particular, may be an important mechanism for explaining differences in child wellbeing across socioeconomic statuses. These differences in wellbeing can persist throughout the life-course and lead to intergenerational social stratification (Conley, 2001).

Living in a crowded home may affect a child's wellbeing in a variety of ways. The lack of a comfortable, quiet space can lead children to have difficulties studying and reading, affecting their school performance. When space is scarce, different schedules held by household members may disturb children's sleep. The lack of productive sleep can lead to difficulty concentrating during the day and negatively affect mood and behavior. In addition, children in crowded housing have a higher probability of catching illnesses, which can interfere with their daily routine and interrupt their schooling (Edwards *et al.*, 1994, Saegert and Evans, 2003).

Lack of privacy can result in stress, difficult social interactions, and behavioral problems for all household members. Parents in overcrowded homes tend to show less responsive parenting .which may lower parents' participation in parent–teacher organizations at school, monitoring of children's academic performance, and help with children's schoolwork. Children in crowded homes have more behavioral problems in school , which can extend to other social contexts.

Children raised in crowded homes may take their educational, behavioral, and physical health disadvantages with them throughout their lives. Their poor performance in school decreases their chances to access higher education and socioeconomic attainment. Behavioral problems can lead to difficulties interacting with others. Negative interactions with teachers, parents, and peers during childhood, can lead to future challenges in forming personal and professional networks in adulthood. Physical health problems can interrupt children's daily routines and keep them behind their peers in school and later in the workforce. Ultimately, children growing up in crowded homes are more likely than others to find themselves in a similar situation as their parents, contributing to the intergenerational transmission of social inequality.

Researchers have explored a variety of outcomes and found that housing crowding leads to adult psychological withdrawal, loneliness , poor marital relationships , negative parent–child relations, less-responsive parenting, higher rates of being held back a grade in elementary and middle school, and increased child behavioral problems at school, while other research has shown no effect. One reason for the inconsistency in housing crowding effects is the variety of crowding measures used in surveys, including persons per room, persons per bedroom, and persons per square foot, which makes comparability difficult. In addition, early studies on crowding do not control for the socioeconomic status of individuals, thereby confounding the effect of crowding and poverty on the outcomes.

Poor housing conditions have significant effects on different aspects of a child's life. These negative effects during childhood can persist throughout life, ultimately affecting their future socioeconomic status and, likely, their adult wellbeing. It is important to identify aspects of a child's living environment that may prove harmful in order to prevent them. If housing agencies and communities are provided with information on the deleterious effects of housing crowding, they can design housing programs that mitigate the effects of crowding and form standards for appropriate household unit size. The living environment, net of socioeconomic status, is an area that can contribute to the intergenerational transmission of social inequality. By better understanding the role of housing in the lives of children, we can ensure a healthy living space and reduce educational, behavioral and physical health disparities.

Toxins and pollutants

Lead

Needleman *et al.* (1979) documented the impacts of lead exposure on young (grade school) children's IQ and externalizing behaviors and cognitive functioning of children and adolescents and other confounding factors.

More recently, a significant body of research has documented the effects of prenatal and childhood exposure to lead on children's current and prospective developmental functioning in middle-income, newly industrial countries such as China, India, the Philippines and Malaysia. In fact, it is estimated that around 40% of children living in economically developing countries have elevated blood lead levels. In addition, some of the most recent work across the globe has found that even very low levels of lead exposure can be toxic to infants and young children. An important characteristic of many toxins is that even after emissions are eliminated, they remain in the ecosystem for a very long time (Meyer, Brown, & Falk, 2008). Lead used to be incorporated in plumbing , and thus can potentially leach into water supplies. Many toys and other common household products used to be made with lead, a practice that unfortunately continues today in China, for example. Another pathway that is perhaps more insidious is the cross-generational transmission of toxins. Some toxins are lipophilic, which means they can be stored in body fat. Thus prior exposure to some toxins, even preconception, can eventually affect the developing organism. Finally, even when children themselves are not exposed to toxins, they may be susceptible to indirect exposure via parental exposures. A common example of this is from agricultural workers who absorb pesticides into their skin and/or their clothes. Tragically, child laborers in many parts of the world remain in direct contact with toxins in agricultural, construction, and manufacturing sectors. Numerous studies document a dose-response function between body lead level burdens and IQ reductions. For example, Canfield *et al.* (2003) found that, after controlling for socioeconomic status and other demographic variables, three- to five-year-olds' blood lead levels were significantly negatively associated with IQ, even at levels of exposure below the US regulated 10 μg/dl level. Teachers reported more attentional problems-deficits in attention, reaction time, and visual-motor integration among elementary school aged children.

In the global South, most of the work to date has considered the impacts of lead on general cognitive functioning, children under 12 years had blood lead levels at or above limit of 10 μg/dl and were diagnosed with cognitive dysfunction. Similar to research in other contexts, a 1-μg/dl increase in blood lead level was associated with a two-point decline in IQ. Similarly, another study found that blood lead levels, statistically controlling for parents' education, household income, and other family demographic factors, predicted children's cognitive functioning. Studies in China across a wide age range have documented similar effects as have recent studies of 6- to 8.5-year-old children in Peru and Ecuador.**Pov:** The evidence for long-term effects of lead on children's cognitive functioning following prenatal exposure is equally strong. Hu *et al.*

(2006) found that maternal blood lead levels in the first trimester, but not in the second or third trimester, predicted 12- and 24-month-old Mexican infants' general cognitive functioning (Mental Development Index, MDI). These effects were large; a 1-SD increase in first-trimester maternal plasma lead level was associated with a 3.5-point decrease on the MDI. Shen *et al.* (1998) similarly found that, after statistically controlling for confounding factors such as family SES and parental exposure to lead at work, 3-, 6-, and 12-month-old Shanghai infants with high umbilical cord lead levels received significantly lower MDI scores than those with lower lead levels. Patel and colleagues (2006) found that the cord blood lead levels of neonates living in Nagpur, India, significantly predicted autonomic stability and abnormal reflexes. In addition, amongst infants with cord blood lead levels of 5–10 μg/dL, lead levels significantly predicted arousal state regulation, motor functioning and autonomic stability. These findings suggest that lead exposure has important effects on early motor functioning, even at very low levels. Several other studies have shown linkages between early lead exposure and impulsivity, aggression, and hyperactivity in children (Chandramouli, Steer, Ellis, & Emond, 2009; Chiodo *et al.*, 2004; Evans, 2006; Hubbs-Tait *et al.*, 2005).

However, in an early study of the impacts of prenatal lead exposure on both cognitive and socioemotional functioning at ages 2, 4 and 7 years in Kosovo, Factor-Livak *et al.* (1999) found that children's behavior problems were associated with blood lead levels which were positively associated with passive off-task behaviors within classroom settings and negatively associated with activity levels during recess.

Mercury

The impacts of exposure to mercury on children's cognitive functioning are well documented. Low-level maternal mercury exposure damages infant sensorimotor functioning and IQ scores and language development. In addition, high-level maternal mercury exposure has been reported to adversely affect cognitive and physical prenatal and neonatal development. Two major longitudinal projects, one in the Seychelles and one in the Faroe Islands have documented the adverse impacts of prenatal exposure to mercury from maternal consumption of seafood on young children's cognitive functioning. Nevertheless, together these projects suggest that young children's motor and cognitive development, and language, attention and memory in particular, are compromised following prenatal exposure to methyl mercury.

Polychlorinated biphenyls (PCBs)

Prenatal exposure to polychlorinated biphenyls (PCBs), which are used in the manufacture of vinyl and other plastic compounds, has been linked with children's cognitive and socioemotional functioning. These compounds have been banned in most high-income countries, but they continue to persist in environments across the globe, particularly as they tend to bioaccumulate in fish and other animals. Prenatal PCB exposure due to fish ingestion from polluted lakes has consistent adverse effects on neonatal developmental status (especially hyporesponsiveness) and memory among preschool and elementary school aged children. One other developmental aspect of toxin exposure and children's maturation worth mentioning is that lower SES contexts appear to accentuate the harmful impacts of toxins on children's development (Evans, 2006). This might occur for several reasons, including chronic stress, levels of cognitive stimulation in the home, co-occurrence of other toxin exposures, co-occurrence of other risk factors, and, for older children, poorer quality school environments.

Pesticides

Research on the developmental impacts of direct residential pesticide exposure or indirect prenatal or occupational exposure (on the skin or clothing of exposed caregivers) is somewhat limited. However, there is an extensive research literature documenting severe impacts of pesticide exposure on both rats and in vitro models of the mammalian brain. Since pesticides are neurotoxic agents, they may well have serious effects on the developing brain. Indeed, in a recent review, Jurewicz and Hanke (2008) conclude that there is good evidence for the impact of various pesticides on motor functioning (abnormal reflexes) in the newborn human and both motor and cognitive functioning (particularly reaction times, attention and short-term memory) on children. The developing fetus and young children have lower levels of the detoxifying enzymes that may deactivate organophosphate compounds in adults (Furlong *et al.*, 2006). This suggests that the effects of agricultural pesticides on children may be particularly problematic. Few find significant relationships between cord blood and maternal serum levels of DDE, DDT and related compounds on the cognitive and psychomotor functioning of both infants and young children were observed. More contemporary organophosphate pesticides may similarly impact reflexes in infants, reaction times in early childhood, and infant and early childhood psychomotor development and effects on specific cognitive skills, particularly short-term memory and attention. DDT and DDE are currently commonly used in the global South , yet there is almost no research documenting the impacts of these compounds on children's developmental functioning. Children's current exposure was negatively associated with reaction times, but not with other cognitive measures. In a study using a similar design, comparing children living in rural areas with high pesticide use to those residing in low pesticide use areas, 4- to 5-year-old Indian children showed a similar profile (Kuruganti, 2005). And Rodríguez (2012) found that 7- to 9-year-old children of Nicaraguan agricultural workers who were exposed to a variety of pesticides prenatally had deficits in working memory, verbal comprehension, and overall IQ. Thus there is some evidence that prenatal exposure may be particularly problematic, but that later exposure may also impact some aspects of children's cognitive development. The research documenting effects of pesticide exposure on children's socioemotional development is limited, and the findings are mixed (Ruckart *et al.*, 2004). These findings make sense, given other results documenting the impacts of pesticide exposure on children's attention processes. Clearly, more research on the impacts of pesticide exposure on the socioemotional functioning of young children is warranted.

Air pollution

With ongoing rapid industrialization and urban growth, poor air quality is a serious concern in much of the global South, as well as in newly industrial countries in general (Bartlett, Hart, Satterthwaite, de la Barra, & Missair, 1999). Among the most common pollutants to be studied for its effect on cognition is nitrogen dioxide (NO2), a toxicant produced by fossil fuel combustion and thus closely linked to road traffic as well as gas stoves which is found to be associated with poor performance on neurobehavioral tests and memory evaluations, while no similar correlation was found between NO2 exposure at school and cognitive outcomes (van Kempen *et al.*, 2012).

Prenatal exposure to environmental tobacco smoke is negatively associated with cognitive performance at age two in African American and Dominican children in New York City, exposure to high levels of airborne polycyclic aromatic hydrocarbons (PAH) was associated with lower cognitive scores and moderate developmental delay at age three and lower IQ scores lower non-verbal IQ scores among five-year-olds There are also potentially prolonged consequences of overexposure to PAHs. Noting that children highly exposed to PAHs were 2.89 times as likely to have lower MDI scores than unexposed children at the age of three, Perera *et al.* (2006) suggested that greater exposure to such high levels of pollution could adversely affect language, reading and math abilities later on. Changes in brain structure as a result of exposure to high levels of air pollution have been proposed as a possible explanation for resulting cognitive defects while a moderate negative correlation between indoor open-fire cooking and block building performance, memory, pattern recognition and structured play was found.

Water pollution, sanitation and access

Many families in the global South have limited access to clean water and sanitation facilities. The most common water pollutant studied in relation to children's development is arsenic. In a study , 6- to 8-year-old children, those with higher concentrations of urinary arsenic performed worse on several measures of cognitive and language development than did children with lower concentrations. This relationship was not impacted by lead exposure, demographics, or nutritional factors, although lower SES children had higher levels of urinary arsenic, later leading to lower scores than unexposed adolescents on cognitive assessments of memory and attention switching, even after controlling for education and gender. And in a study of 9.5- to 10.5-year-old children using tubewells in Bangladesh, found that water arsenic levels were associated with poorer cognitive functioning and lower mathematics scores than those not exposed, even when controlling for schooling history, prior arsenic exposure, and parental factors. High manganese levels in the public water system may also impact children's behavior. Hair manganese was significantly associated with hyperactivity and oppositional behavior, as measured by teachers' report. Interestingly, the positive relationship between hair manganese and hyperactivity was greater for older children (above 11 years old). Besides their health effects, water-borne diseases also have adverse developmental consequences and cognitive functioning. Likewise, in an investigation of the environmental conditions (including poor access to drinking water, an inconsistent electricity supply and inadequate sewage drains) impacting 7- to 8-year-old children's cognitive development in war-torn Baghdad City, Ghazi and colleagues (2012) found that below average water quality (as reported by parents) was associated with lower IQ scores, and that access to services (including water quality, electricity supply and access to grocery stores) independently predicted IQ, after adjusting for parent education and income. In addition to direct impacts on cognitive functioning, diarrhea and intestinal parasites resulting from bacteria-contaminated water (often from sewage) contribute towards malnutrition and stunting, both of which impact children's IQ and school performance, and may also contribute towards behavioral problems (Bartlett, 2003). These associations may result as early malnutrition and exposure to environmental toxins and stress can alter both brain structure and function, thus leading to long-term changes in cognitive and socioemotional functioning (Grantham-McGregor *et al.*, 2007). In addition, both illness and malnutrition may lead to increased absences from school and attention problems when in school. Finally, it is worth noting that global climate change is likely to affect access to clean water for millions of low-income families in the global South, particularly in Africa and parts of Asia, in the next 20 years (Bartlett, 2008).

Desertification

Desertification is a dynamic process , not necessarily exclusive of any particular region, Erosion, salinization, waterlogging can occur in different types of land. It reduces the sustainable base of society at large. Right now, however desertification is already a disaster for those populations directly affected, living in the most deprived areas of the world. They have lost their livelihood and suffered a sharp deterioration in their nutritional and health status. Whole communities have also been deprived of their economic activities; their social institutions have been disrupted and their most producting members have had to mitigate and seek relocation elsewhere.

Desertification include two phenomenon: Arid land and Drought which are the causes for temporary desertification. It affects irrigated rainfed land and rangelands which are degenerated (Decrease in the herbaceous cover)

Desertification is a socio-economic and natural process which reduced the fertility and biological productivity of the soil to the level which characterizes deserts. It is one of the clearest results of the close and increasing interrelationships between social and natural systems. Though desertification affects mainly arid and semi arid areas, it is also found elsewhere. It is the result of long historical process through which natural phenomena and human activities reinforce each other in changing the characteristics of natural environments.

Desertification has become a natural and social disaster with many humanitarian aspects that require adequate attention and prompt action. A deteriorating natural environment and the dire poverty of the population in the areas at risk also reflect the failure of the international community to prevent and control desertification.

The socioeconomic causes of desertification are related to such inadequate land use practices as over cultivation, erosion, over exploited, overgrazing, woodland clearance and mismanagement of water resources. The combined effects of deforestation, water logging, erosion etc will lead to fall in the water table and to salinization problems. The misuse of wells- frequent where nomads are resettled-also compound difficulties. The population density of arid lands is relatively low. Declining birth rate in arid lands has also been attributed to temporary female infertility because of malnutrition and absence of spouse. The long lasting drought combining with the long term desertification process has resulted in massive migration.

Additional factors in relation to the rate of population growth in arid lands refer to high mortality rates which do not necessarily take place during drought but rather during the rains after the end of the drought and as consequences of diseases like pneumonia, malaria and TB.

The immediate consequences of desertification is to increase the threat to health and well being.

Lower productivity of the natural system means that the land is no longer capable of sustaining the livelihood of the population. The immediate consequences are deteriorating nutritional status and increased morbidity and mortality, especially among the more vulnerable groups such as children and the elderly.

The ultimate response to desertification in migration. In the abandoned area (Outmigrant area) migration implies the destruction of family patterns. This disrupts population dynamics but also entails the loss of the more innovative individuals. Declining land productivity, increasing labour shortage caused by migration, and the fact that those who stay behind (women)have to devote more to survival activities (Fetching water and firewood) cause a further fall in agriculture productivity and in overall economic activity. The traditional institutional and social fabric has been destroyed by environmental deterioration and desertification. In developing countries reduction in agri land, more intensive cultivation, greater use of fertilizers. In less developed countries the penetration of technology associated with economic schemes ill suited to the region has contributed to further environmental deterioration and social disruption.

Increased urbanization also indirectly contribute to desertification. African and Asian city dwellers cook and heat themselves with charcoal and therefore put pressure on local wood resources

Increasing urbanization also forces govt to adopt food pricing policies with a distinct urban bias, which penalize rural producers depriving them of all incentive to increase production.

Desertification hits poorest of the poor, those who are least able to cope. The victims of desertification are those with little or no land, no political power and tiny incomes. For them survival depends critically on the crops, animals and trees which provide their basic needs. Hunger and disease arrive quickly in the wake of any external disturbance, such as drought. ()women inBurkina areas had to walk 6 hr to fetch firewood thrice a week.

The scarcity of energy to satisfy their basic needs results in additional heavy and time consuming work as well as changes in nutritional patterns. More efforts goes to producing less food. Families only have one cooked meal a day and a number of traditional stapled which take longer to cook and require more fuel are often replaced by food of lower energy and protein value.

Things are made worse by the lack of adequate substitutes in dryland, less nourishing food leads to malnutrition which in turn increases the vulnerability of people to different diseases and finally reduces their productive capacity.

The reduction of female productive activities is made worse by the emigration of men or complete families, a new cause of decreasing food production, for the family immigrations brings structural change as women take n the new role of head of the family, who are BPL. Changes in the individual and family level tend to cause changes in society as a whole which also has to contend with the penetration of new habits and new technology in the form of schemes as ranching, mechanization, cash crops etc.

These changes associated with long term ecological deterioration and the occasional drought or the disasters deal a death blow to the institutional and control mechanisms developed over the centuries or at any rate make them useless.

Noise

Numerous studies in high-income countries reveal that chronic noise exposure early in childhood interferes with reading acquisition (Evans, 2006). Adverse impacts on reading have also been replicated in prospective longitudinal studies with the introduction of a new major noise source such as an airport, as well as in experiments with noise attenuation interventions. Children in higher elementary school grades suffer greater adverse reading outcomes; this has been attributed to longer duration of exposure but might also reflect greater awareness of noise. Some studies have also shown worse reading outcomes for children exposed to noise at home and school, bolstering the duration of exposure explanation. Children with poorer cognitive skills also appear more vulnerable to the induction of reading deficits from noise exposure . Several cognitive deficits reliably associated with noise exposure are candidate mechanisms for the well-documented noise-reading link. Long-term memory is adversely affected by noise, and attentional strategies are altered by noise exposure. Interestingly, a few studies have also shown linkages between chronic noise exposure and deficits in auditory discrimination (e.g., phoneme perception), a critical aspect of speech perception. Speech perception is a major building block of reading acquisition. Finally, emerging work in neuroscience indicates potentially detrimental noise effects on brain speech function and structure (Kujala & Brattico, 2009). Chronic noise exposure, similar to many of the environmental conditions described herein, is not only aversive but also uncontrollable and sometimes unpredictable as well. Repeated exposures to uncontrollable as well as unpredictable events can undermine human motivation, thus impacting the persistence and effort needed (amongst other things) for academic achievement. The bioecological perspective (Bronfenbrenner & Morris, 1998) suggests a complementary set of processes that might also be related to noise and reading acquisition. Noise might alter caregiving behaviors salient to reading acquisition. We know, for example, that teachers in high noise impact schools alter their teaching methods and also complain about interruption and fatigue. It is conceivable that parents might talk less to their children, be less responsive to children's verbalizations, and not read aloud as much to their children in high noise settings. Outside of the global South, Hiramatsu and colleagues (2004) found deficits in long-term but not short-term memory among 8- to 11-year-olds residing proximate to a large air force base in Okinawa, Japan compared to their peers living in quiet areas. Children in the noisier schools have decreased social skills and more oppositional behaviors but were similar in levels of anxiety compared to their peers attending quieter school as shown a link between chronic noise exposure and elevated learned helplessness among children (Evans, 2006)

Crowding

The most consistent crowding metric with human consequences is people per room. Indices of external density such as people per census tract typically yield no associations with human behavior (Evans, 2006). Studies that have teased apart residential density from family size find the former rather than the latter to be the more critical variable. Although many believe there are differences in tolerance for crowding across different cultural contexts, the cognitive and behavioral development of children living in as diverse contexts as the United States, India, Thailand, Egypt, Hong Kong, South Africa, and Jamaica indicates similar developmental correlates of crowding in both residential and school settings. It is important to note that children in the global South, relative to North America and Western Europe, tend to live in more crowded home environments. Significant research across multiple contexts documents the impacts of crowding on general school achievement and IQ, reading comprehension, and object spatial relations. Infants studies also reveal that, infants in more crowded households were more lethargic and drowsy, conditions associated with delayed development. These associations may result from disruptions of children's exploration, play and engagement with both objects and people in their immediate environments. Crowding in educational environments has also been linked to more off-task time. Children from more crowded homes spent less time engaged in play with objects, more time unoccupied, and more time as onlookers. Residential crowding can also disrupt parent-child interactions, in more crowded homes, parents talk less with their infants and toddlers and use less complicated vocabulary and sentence structures with their toddlers . Children's physical development and quantitative skills and academic achievement were also adversely associated with household crowding, due to heightened family conflict and residential conflict. More crowded apartments also have inadequate privacy for students to study. Crowded home and school environments significantly impact the behavior and socioemotional functioning of both children and their parents. Interestingly, in a study of feeding practices in Jamaican primary schools, Grantham-McGregor, Chang, Walker, and Powell (1998) found that the negative impacts of classroom crowding on children's behavior were exacerbated by poor nutrition. One of the effects of high-density living may be greater difficulty monitoring and regulating children's behaviors. Less parental monitoring is a well-documented predictor of behavioral conduct disorders, including juvenile delinquency. In high-income countries both children and parents report more strained, negative familial interactions in high-density homes, as well as instances of elevated punitive parenting practices. Children in more crowded preschools and elementary schools also evidence more aggressive behaviors towards their classmates One of the factors believed to drive part of the crowding – aggression link is conflict over scarce resources such as toys. One of the ways in which crowded family members appear to cope with crowding is to socially withdraw from one another, which can have the unintended consequence of diminishing socially supportive relationships. A number of studies, including some with random assignment, have shown that crowded children tend to be more **socially withdrawn, parents are less responsive**. Given greater social withdrawal among children in high-density homes and lower levels of parental responsiveness in similar situations, some investigators have explored whether crowding might also be linked to psychological distress among children. **As indicated above, there is already evidence of elevated rates of aggression, withdrawal, and behavioral conduct disorders such as juvenile delinquency**. Family conflict partially mediated these relationships. There is also evidence that, resting blood pressure was elevated among more crowded boys, but not girls. This matches several studies indicating elevated indices of physiological stress among children living in more crowded homes or attending more crowded schools/childcare.

Residential mobility

Poverty, substandard housing, and slum dwellings without security of legal tenure often lead to excessive residential mobility. Reliable housing is critical for children's security and stability, and is essential if families are to establish daily routines (Bartlett *et al.*, 1999). High levels of residential mobility are associated with **poorer psychological adjustment, less socially supportive peer relationships, and deficits in academic achievement.** In addition, students and teachers in classes with high levels of mobility face considerable challenges because of the instability of their members. Early childhood residential instability can also influence developmental trajectories. Adolescents with more frequent moves tend to have diminished social networks and hold comparatively less central positions therein , and are vulnerable to earlier onset of sexual activity . In the global South, residential mobility is high, particularly for low-income families living in urban areas (Bartlett *et al.*, 1999), who frequently face forced evictions (Chatterjee, 2007). Although little work in the global South has directly evaluated the impacts of high residential mobility on children's cognitive and socioemotional functioning, it is likely that high mobility disrupts proximal processes (Bronfenbrenner & Evans, 2000). Further, children whose families are evicted from their homes in a violent manner may experience trauma.

Housing type

Research on housing type in more affluent countries has focused primarily on the potential developmental implications of high-rise housing. Several studies in high-income countries have found that children and youth in high-rise buildings manifest greater levels of behavioral conduct disorders (e.g., delinquency, aggression). In a study of families living in high- versus lowrise apartments in Israel, Churchman and Ginsberg (1984) similarly found that the outdoor play behavior of **4- to 5-year-old** children living in high-rises was more restricted than that of other children, although it should be noted that these effects were not found at other ages (within the range of 2–13 years). In the global South, housing type is inextricably connected to housing quality. In addition, informal housing is typically unstable, and children living in such areas frequently face eviction and therefore frequent residential mobility . In addition, children living in informal housing may be more vulnerable to injury, and are more likely to be exposed to toxins from industrial waste. And children who are homeless or who live in informal housing may be less likely to attend school, as they lack a formal address (Wegelin & Borgman, 1995). For example, a recent survey in Delhi found that only 54.5% of children in slums enrolled in school, as compared to 90% across the city as a whole (Aggarwal & Chugh, 2003). For those in school, homelessness has significant impacts on school performance and socioemotional well being (Hicks-Coolick *et al.*, 2003; Neil & Fopp, 1992). There is also some evidence that children's self-esteem is negatively impacted by residence in slum dwellings and other informal settlements (Kruger, 2002). Delays in cognitive development associated with residential density among preschool children are attenuated if children have access to a room where they can spend time alone (Wachs & Gruen, 1982).

Resources for children

Another aspect of the physical environment that may influence young children's development is the availability of learning materials (Bradley & Corwyn, 2005; Bradley & Putnick, 2012). However, the availability of such materials is seldom disentangled from parent-child interactions in the literature. Nevertheless, there is a strong relation between income and the provision of both stimulating materials and experiences for young children from birth through adolescence (e.g., Bradley, Burchinal & Casey, 2001; Evans, 2004; McLoyd, 1998). And, several studies have shown that cognitive enrichment in the home mediates much of the co-variation between parental income and child cognitive development. Access to other material resources such as electricity, a radio, a television, a telephone and transportation may also impact children's cognitive development in particular. Access to resources was positively correlated with children's general cognitive functioning, perspective taking, and levels of exploratory play.

Schools and childcare

There is a significant body of research investigating the impacts of school quality on children's school achievement . However, as is true for the work on home environments, little research has specifically investigated the impacts of the physical environment of schools on children's developmental outcomes, particularly in the global South. Most research in the US and Europe on the physical characteristics of educational settings has focused on open versus traditional plan configurations (Evans, 2006). Because this issue has tangential relevance at best to children throughout most of the world.

School and classroom size

There is a large body of research on school and classroom size. Because nearly all of this work has been conducted within the US and Western Europe, we do not know what happens when much larger scale schools or bigger classrooms occur. Although there is some variation across regions, primary school pupil-teacher ratios (PTRs) in the global South are typically much higher than those in the global North. Students in smaller schools in the US and Western Europe perform slightly better on standardized tests and feel more connected to their school. There is some evidence that the benefits of smaller school size are greater for low-income children, and for children in lower grades. Similarly, classroom size research yields a relatively consistent picture of small, adverse effects on children in both high- and low-income countries with increasing size. There is also some evidence that smaller classrooms support more student- as opposed to teacher-directed learning and, similar to school size, are associated with more socially supportive settings.

Physical quality

In the global South, the majority of rural schools in particular have inadequate building facilities, including a lack of finished flooring (Glewwe *et al.*, 2011; Riddell, 2008). In many countries, half to two thirds of schools lack electricity, water, and basic sanitation facilities (UNICEF, 2010). One important limitation in most work on educational settings and student achievement, however, is over-reliance on school professionals' ratings of building quality. Since teachers and administrators are well aware of children's achievement profiles in their own schools and are themselves likely affected by building quality, the potential for spurious associations in this measurement approach is considerable. However, assessments of building quality conducted by independent raters (e.g., structural engineers) have also been consistently associated with standardized test scores. Improvements in the physical structure of schools in the global South do appear to positively impact students' test scores . However, the research to date in this area is very tentative, and typically the schools being compared have multiple factors that differ in quality, making it difficult to clearly identify individual influences on children's outcomes. In a recent meta-analysis of the research to date on the impact of school quality, including both physical and psychosocial factors, on children's school achievement in low-income countries, Glewwe *et al.* (2011) found that there appears to be good evidence for the impact of access to electricity on children's educational outcomes. In high-income countries, where lighting is typically sufficient, research has focused more on potential benefits of exposure to natural light. In North America, upper respiratory infections, asthma and allergies are the most common cause of primary school absenteeism and have been routinely linked to exposure to mold and other allergens as well as ambient pollutants inside both schools and children's homes (EPA, 2003). Poorly maintained heating and ventilation systems as well as low levels of indoor: outdoor air exchange exacerbate these adverse indoor climate impacts on children (Evans, 2006). Consistent with the bioecological perspective (Bronfenbrenner & Morris, 1998), in addition to focusing on the direct effects of school setting physical conditions on children themselves, it is important to keep in mind that substandard working conditions influence labor satisfaction and retention, and the same holds true for teachers. Several studies have shown that poor quality school physical conditions adversely influence teacher satisfaction and retention (Buckley *et al.*, 2004). Resources In the global South, there is some evidence that access to basic resources in school environments, such as a sufficient number of desks, tables and chairs; access to blackboards; access to textbooks and other books; and the availability of a school library all impact children's school achievement. From the above discussion, it may be inferred that below mentioned are the major Economic Factors that Affect Human Growth and Development

Economic Factors that Affect Human Growth and Development

Economic development and human development are related but distinct concepts often used to assess a country's progress and well-being. In the case of India, both dimensions have seen significant advancements, but challenges remain. Economic development primarily focuses on the growth of a nation's economy and the improvement of material well-being. Human development goes beyond economic indicators and encompasses broader aspects of well-being, including health, education, and quality of life. The disparity between human development and economic development in India is a complex issue with multiple contributing factors. While India has experienced significant economic growth in recent decades, improvements in human development indicators, such as education, healthcare, and living standards, have not kept pace.

Inequality

India grapples with high levels of income and wealth inequality. Economic growth has disproportionately benefited certain segments of the population, leading to a widening wealth gap. This inequality can limit access to quality education, healthcare, and other essential services for a large portion of the population.

Healthcare Disparities: Lower income levels might result in poor access to healthcare, which can lead to untreated illnesses affecting physical and mental health. On the other hand, good economic standing can afford high-quality healthcare, regular check-ups and preventive measures, fostering good health. Despite economic growth, access to quality healthcare remains limited in many parts of India, especially in rural areas. High out-of-pocket healthcare expenses can push families into poverty, and preventable diseases continue to affect a significant portion of the population.

Education: Economic stability can significantly influence a person's level of education. A higher level of education often leads to better jobs and greater economic stability. Early childhood education is fundamental for cognitive development and aids future economic status. While India has made strides in increasing literacy rates, the quality of education remains a concern. A large number of children attend schools with inadequate infrastructure, underqualified teachers, and outdated curricula. This affects the overall development and employability of the workforce.

Gender Disparities: Gender inequality continues to hinder human development in India. Discrimination against women and girls in terms of access to education, healthcare, and employment opportunities limits their contribution to the workforce and overall development.

Malnutrition and Food Security: Malnutrition remains a critical issue in India, particularly among children. Poor nutrition during the early years of life can have long-term impacts on physical and cognitive development. Nutrition has a direct correlation with income levels. Lower income might mean less access to healthy and nutritious food affecting physical growth and immunity. Poor nutrition can lead to obesity, malnutrition and other health-related issues which can impact growth and development.

Sanitation and Clean Water: Access to clean drinking water and sanitation facilities is uneven across India. Poor sanitation and lack of access to clean water can lead to health issues and hinder human development.

Social and Cultural Factors: Deep-rooted cultural norms and social hierarchies can perpetuate disparities in access to education, healthcare, and economic opportunities. Discrimination based on caste, ethnicity, and religion can limit the development prospects of marginalized groups.

Regional Disparities: India is a vast and diverse country with significant regional disparities in human development. States in the southern and western regions tend to have better human development indicators compared to states in the northern and eastern regions.

Government Priorities: The allocation of government resources and policy priorities can affect human development outcomes. Shifts in public spending and policy focus are necessary to address disparities.

Infrastructure Gaps: Inadequate infrastructure, including transportation, electricity, and digital connectivity, can limit economic opportunities and access to services in rural and remote areas.

Income and Poverty: The income of a family or individual has a substantial influence on the opportunities and resources available, affecting overall health and wellbeing. Those with lower incomes might not have access to sufficient healthy food or safe housing, negatively impacting physical growth and health. Poverty can lead to chronic stress, which can affect mental and emotional development.

Housing: Economic factors directly influence the quality and stability of housing. Poor housing conditions can expose an individual to health hazards and negatively affect physical health. Unstable housing can lead to chronic stress, affecting emotional and cognitive development.

Employment: Having stable, good quality employment can provide economic stability, reduce stress levels and offer better living conditions. Employment also provides socialisation opportunities, impacts self-esteem and contributes to one's mental and emotional wellbeing. Long-term unemployment, however, can create chronic stress and result in poor mental health.

3

Socio-Cultural Ecology and Human Development

3.1 Physical ecological factors, Terrain, climatic changes, demographics and their effect on human development

As we delve deeper into the world of dyads, certain key concepts emerge as the cornerstones of our understanding. These ideas are like the primary colors on an artist's palette – fundamental elements that combine to create the rich, complex picture of dyadic relationships.

Reciprocity and mutuality are at the heart of many dyadic interactions. It's the give-and-take, the back-and-forth that characterizes healthy relationships. Like a well-choreographed dance, partners in a dyad often mirror each other's actions and emotions, creating a sense of balance and harmony.

Power dynamics, on the other hand, can sometimes throw that dance off-kilter. In any relationship, there's often an ebb and flow of influence and control. Sometimes it's obvious, like in a boss-employee relationship. Other times, it's more subtle, shifting based on the situation or topic at hand. Understanding these power dynamics is crucial for navigating relationships successfully.

- **Communication** patterns are the lifeblood of dyadic relationships. It's not just about what's said, but how it's said, when it's said, and even what's left unsaid. Some couples finish each other's sentences, while others struggle to find common ground. These patterns can be a window into the overall health and functionality of the relationship.
- **Conflict resolution** is another key concept in dyadic psychology. Let's face it – no relationship is without its bumps and bruises. How couples navigate disagreements can make or break a relationship. Some fight like cats and dogs but make up quickly, while others avoid conflict altogether. Neither extreme is ideal, and finding a healthy middle ground is often the key to long-term relationship satisfaction.
- **Emotional contagion** is a fascinating phenomenon in dyads. It's the idea that emotions can spread from one person to another, like a yawn in a crowded room. Have you ever noticed how your mood can shift just

by being around someone who's particularly cheerful or grumpy? That's emotional contagion at work, and it plays a significant role in dyadic interactions.

The most crucial interpersonal context for the developing infant is the small number of care giving relationships that the child encounters. Through these relationships the infant begins to understand his world, learns how to interact with others, and begins to develop a sense of his competence and self-worth. After all, the impact of infant's experiences of environmental risk factors, such as poverty, maternal mental illness, and partner violence is primarily via their effects on the infant–parent relationship. Further, intrinsic risk factors, such as biological difficulties, are moderated by the infant–parent relationship. For example, infants with complications of prematurity have better outcomes when their caregiving environments are supportive, and more problematic outcomes when their caregiving environments are less supportive. Also, difficult temperaments can be moderated through a responsive, nurturing, and consistent caregiving experience.

The attachment relationship is a biologically based process that motivates the young child to seek comfort, support, nurturance, and protection in times of distress from discriminated attachment figures – providing the basis for psychological security as well as physical safety. The attachment relationship develops over the course of the first year of life through the myriad of daily interactions between the infant and the primary caregiver(s). The quality of the infant–caregiver relationship is a risk or protective factor for infants' later development. **A warm, nurturing, sensitive, responsive, and consistent pattern** of interactions between the infant and caregiver leads to a 'secure' attachment; through these interactions, the infant learns that he is worthy of being taken care of, that he can count on his caregiver to be there when he needs her, and he develops a sense of self-competence in that his actions (i.e., signals, cues, behaviors, communications) can be understood and are effective in getting his needs met. Conversely, interactions that lack these positive qualities and are inconsistent, unpredictable, harsh, or punitive lead to insecure or disorganized attachments. Preferred attachment appears in the latter part of the first year of life, heralded by the appearance of separation protest and stranger wariness. Infants become attached to caregivers with whom they have a significant number of interactions. If more than one attachment figure is present, infants develop a hierarchy of preferred caregivers to whom they turn for comfort, support, nurturance, and protection.

During these early months, the infant appears to develop a set of expectations, termed 'working models', of what it is like to be in an intimate relationship with another person. These models are relationship specific, so that the infant's experiences with each caregiver determines the nature of the expectations that the infant develops for his or her relationship with each caregiver. This relationship specificity has significant implications for assessment and treatment.

A salient objective feature of the social environment in which people find themselves is group size. Knowledge of group size is highly relevant to behavioural scientists given that humans spend considerable time in social settings and the number of others influences much of human behaviour.

Our thoughts, emotions, and behaviours are influenced by those around us– in conversations at home, in shared activities such as going out for dinner with friends, and while working on projects with colleagues. In fact, humans spend considerable time in social settings. Research finds that the *number of others* around us is key to understanding social phenomena such as free riding, trust, or helping behaviour. For example, people are less likely to free ride in smaller groups, more prone to trust others in smaller groups, and they are more inclined to feel responsible to help others in need when part of a small group, compared to larger group sizes. A dyad refers to two individuals treated as a unit, or a pair. Dyadic relationships are characterized by some kind of interdependence, such as romantic relationship, friendship, collaboration, or social interaction.

A relation obtains whenever one person in a setting pays attention to or participates in the activities of another. The presence of a relation in both directions establishes the minimal and defining conditions for the existence of dyad. A dyad is formed whenever two persons pay attention to or participate in one another's activities. The dyad is important for development in two respects. First, it constitutes a critical context for development in its own right, second, it serves as the basic building block of the micro system, making possible the formation of larger interpersonal structure-triads, tetrads, and so on.

Theoretical Frameworks: The Lenses Through Which We View Dyads

Bronfenbrenner's Ecological Systems Theory outlines several environmental systems, each representing different levels of influence on an individual's development. While the theory itself does not explicitly describe "functional forms that a dyad can make," we can interpret and apply his systems to understand how dyadic relationships might be influenced or characterized within these broader contexts:

Microsystem: This is the closest, most immediate layer of the ecological systems, encompassing the direct interactions an individual has with their environment, such as with family, school, or peers. In terms of dyadic relationships, the microsystem would involve direct, face-to-face interactions and relationships, such as those between parent and child, romantic partners, or close friends. These relationships are deeply personal and have a significant impact on an individual's development and behavior.

Mesosystem: This layer refers to the connections between the microsystems, or how different parts of a person's life interact. For dyads, the mesosystem might involve how their relationships are influenced by or influence other aspects of their environments. For example, the relationship between a teacher and student (a dyad in the microsystem) might be affected by the school's culture or the student's family life, showing how multiple contexts intersect and influence each other.

Exosystem: This layer encompasses the external environmental settings that indirectly affect the individual, even if the individual does not actively participate in them. In terms of dyadic relationships, this could involve how broader social, economic, or political conditions affect the dynamics between two people. For example, the stress of a parent's job loss (an exosystem factor) could influence their relationship with their child, even though the child is not directly involved in the parent's workplace.

While Bronfenbrenner's theory is more focused on individual development within these various ecological systems, it can be useful in understanding how dyadic relationships are influenced by and exist within broader social and environmental contexts. Each level or system offers a different lens through which to view the dynamics of dyadic relationships, emphasizing the complex interplay between individual, relational, and contextual factors in human ecology.

Social Exchange Theory: This theory suggests that relationships are like a marketplace where we trade social goods – love, support, companionship – and try to maximize our benefits while minimizing costs. It's not as cold and calculating as it sounds, though! This framework helps us understand how people make decisions within relationships and why some partnerships thrive while others fizzle out.

Attachment Theory: the brain child of John Bowlby and Mary Ainsworth. This theory is like the origin story of our relational behaviors, tracing our adult relationship patterns back to our early bonds with caregivers. It's fascinating stuff – our childhood experiences create a template for future relationships, influencing everything from how we seek comfort to how we handle conflicts.

Interdependence Theory, examines how partners in a dyad influence each other's outcomes. It's like a relational chess game, where each person's moves affect the other's choices and experiences. This theory helps us understand concepts like commitment, trust, and power dynamics within relationships.

Systems Theory zooms out to look at dyads as part of larger social systems. It's like viewing a relationship as a living organism, with each part influencing and being influenced by the others. This perspective is particularly useful when studying family dynamics or how dyads function within larger groups.

Symbolic Interactionism: This theory focuses on how we create meaning through our interactions with others. It's like learning a new language – in dyads, we develop shared symbols, gestures, and understandings that shape our reality and our relationship.

Interpersonal structures as context of human development

A relation obtains whenever one person in a setting pays attention to or participates in the activities of another. The presence of a relation in both directions establishes the minimal and defining condition for the existence of a dyad- a dyad is formed whenever two persons pay attention to or participate in one another's activities. The dyad is important for development in two respects. First, it constitutes a critical context for development in its own right. Second, it serves as the basic building block of the microsystem, making possible the formation of larger interpersonal structures-triads, tetrads, and so on. In terms of its potential for furthering psychological growth, there are **three different functional forms** that a dyad may take.

Observational Dyad: In this form, one individual observes the behavior, actions, or expressions of another. This can occur in parent-child relationships, where the child learns by observing the parent. It's also common in mentor-mentee dynamics, where observation allows the mentee to absorb knowledge or skills.

Joint Activity Dyad: Here, both individuals actively engage in shared activities or tasks, fostering collaboration and mutual learning. This can be seen in peer interactions, team projects, or even playful activities between a parent and child, promoting cooperation and bonding.

Reciprocal Dyad: This type involves balanced give-and-take, where both individuals contribute equally to the relationship, such as in friendships or partnerships. The mutual exchange of ideas, emotions, or actions helps nurture and strengthen their connection.

Types of Dyads

Dyads can take on various forms depending on the relationship dynamics, goals, and context:

Symmetrical Dyads: Both participants share equal roles and responsibilities. Example: Peer friendships, where both individuals influence each other equally, fostering mutual respect and collaboration.

Complementary Dyads: The roles and contributions of individuals differ but complement each other. Example: A teacher and a student—one provides knowledge, and the other absorbs and applies it. Their roles are interdependent.

Conflictual Dyads: Marked by disagreement or opposing goals. While they may create friction, such dyads can also foster growth when conflicts are resolved constructively. Example: Rivalry between siblings, which can help develop conflict resolution and negotiation skills.

Nature and Function of Joint and Molar Activities

Joint Activities: Involve two people working together on a shared task or experience. Ex: Playing a board game, building something together, or having a conversation to solve a problem. Encourages teamwork, communication, and shared learning.Strengthens emotional bonds through cooperative engagement.

Molar Activities

Molar activities as exhibited by the developing person serve as indicators of the degree and nature of psychological growth; as exhibited by others present in the situation they constitute the principal vehicle for the direct influence of the environment on the developing person. All molar activities are forms of behavior,but not all behaviors are forms of molar activity.The reason for making the distinction lies in the belief that not all behaviors are equally significant as manifestations of or influences on development. Many are so short-lived as to have minimal import; these are referred to as molecular behaviors .Others are more long-lasting but, because they lack meaning to the participants in the setting, have only negligible impact. The definition of molar activity thus emphasizes both some persistence through time and some salience in the phenomenological field of the developing person and of others present in the setting Represent broader, consistent patterns of interactions over time. Examples: A caregiver regularly providing comfort, a mentor guiding a student through their career journey.

DEFINITION: A molar activity is an ongoing behavior possessing a momentum of its own and perceived as having meaning or intent by the participants in the setting.

Function: Builds a foundation of trust and reliability. Influences long-term developmental outcomes by shaping habits and emotional responses.

Affective Relations

The affective domain refers to emotions as well as their outward expression. As with the concept of emotion, descriptions of the affective domain are rather vague, lacking a universal, operationalized definition. While emotion is at the core of the affective domain, it spreads quickly from there. This is because emotion is often seen as involving three subcomponents: feeling, cognition, and behavior. Feeling is the physiological sensation one experiences. Cognition is the subjective thoughts that accompany the sensation. Behavior, which might be facial display, body positioning, or a variety of other actions, is related to both feelings and accompanying cognitions. Thus, the affective domain encompasses physiological, cognitive, and behavioral processes related to emotion. It also encompasses our awareness or discernment of our and other's emotions, the ability to connect our emotions to those of others, the display of emotion, and the ability to manage or regulate one's emotions.

This refers to the emotional aspects of a dyad, including feelings, attitudes, and the depth of connection between individuals.

While the affective domain has been a subject of research for centuries, there are three individuals who are typically considered the founding fathers of research on emotions: Charles Darwin (1998), William James (1884), and Sigmund Freud (1960). Darwin founded his concept of ethology with observations of emotional expression in natural settings and connected them to human evolution. James emphasized physiological changes in the body and showed that emotions are involved in monitoring our bodies. Freud offered the method of listening to what people said about their emotional lives and people may need to discuss their emotions with others in order to be understood. More recently the concept of appraisal has become an important influence on research in the affective domain.

Theories of Emotion

As with any concept, researchers and theorists have differing views as to the function and importance of the affective domain. Some see it as a regulatory system whereas others see it as an activation system. Some see emotion as a precipitating event whereas others see it as a resulting event. This section will discuss some of the current theories and models.

In the Communicative Theory (Oatley & Johnson-Laird, 1995), emotions are viewed as caused by conscious or unconscious cognitive evaluations. Each evaluation produces a signal that is transmitted through multiple processors of cognitive architecture to produce a basic emotion. This signal functions to control organization of the brain in order to ready the mechanisms of action and bodily resources, to direct attention, to set up biases of cognitive processing, and to make the issue the caused the emotions salient in consciousness. The phenomenological experience of the signal is a distinctive feeling, or emotion (e.g. happiness, sadness, anger, etc.). Thus, in essence, emotions are seen as managing goals.

The Feedback Theory (Parkinson & Manstead, 1992) assumes that emotions arise as a consequence of a bodily reaction, rather than cognitive appraisals of a presenting situation. This theory dates to William James (1884), who drew attention to the fact the bodily responses, including facial, postural, motor, and autonomic changes, are central aspects to our idea of experiencing emotion. Our relationship with an object evoking the emotion is expressed through the body (e.g., turn away from unpleasant sights, approach pleasant sights). Thus, part of experiencing emotion is to feel oneself expressing a physical attitude toward an object. Furthermore, James proposed that feelings are a result, rather than cause of emotional behavior (e.g., we are happy because we smile).

A similar theory, the Discrete Emotion Theory (Fogel, Nwokah, Dedo, Messinger, Dickson, Matusov, and Hold, 1992) suggests that emotions organize and motivate action such that a discrete emotion can be defined as a particular set of neural processes that lead to a specific expression and a corresponding specific feeling. The emotion program is believed to be phylogenetically adapted with respect to the basic function of survival. This theory breaks down each emotion, suggesting that patterns of neural stimulation cause associated changes in feeling, and are associated with distinct sets of facial, vocal, respiratory, skin, and muscle responses. These theorists focus primarily on the face in expression of emotion, and believe emotional development is controlled by maturation of the central nervous system (CNS), and that the organism must learn rules that modify and modulate expression. An example of research that supports this position had participants hold a pencil between their teeth for a period of time, a task that uses the facial muscles involved with smiling. Results showed that the participants reported feeling happy.

According to the Functionalist Model (Campos, Mumme, Kermoian, & Campos, 1994), emotion is "the attempt by the person to establish, maintain, change, or terminate the relation between the person and the environment on matters of significance to the person" (p. 285). This model is closely aligned with Lazarus' Relational model (1991) discussed in more detail below. Emotional development begins with a core set of CNS emotions programs, which are defined with respect to basic functional or survival relationships between the individual and environment, and involve tendencies of the entire body. Emotions are part of innate routines for social communication and serve to initiate and maintain contact with others.

Lastly, the Social-Constructivist Model of Emotion (Jenkins *et al.*, 1998), similar to the functionalist model, also views emotional experience as embedded in the conditions that justify it. This perspective emphasizes that we learn to give meaning to our experiences through our social exposure and cognitive developmental capacities. Thus, one's emotional experience is contingent upon specific factors, which contribute to our learning what it means to feel something and then do something about it.

Affective relations can be **positive** expressed through love, trust, and mutual respect. Example: A parent-child relationship characterized by warmth and support fosters security and confidence in the child. Positive affective relations nurture personal growth, emotional stability, and social skills. Affective relations can be **negative** marked by feelings such as anger, resentment, or mistrust. Example: A strained friendship where misunderstandings or competition dominate the interaction. Negative affective relations, if unresolved, may lead to stress but can also spark self-reflection and change.

Balance of Power

Power balance, that is, equal levels of potential influence between relationship partners, has been linked to relationship happiness.

Power and Relationship Quality

Power is typically understood as the ability to influence others on the basis of asymmetric outcome control (Keltner *et al.*, 2003; Overbeck, 2010). The subjectivity of power, that is, the belief that one has the potential to influence others, is often termed felt power or personal sense of power (Anderson *et al.*, 2012; Tost, 2015). Felt power is a central variable in understanding several outcomes within intimate relationships (Kim *et al.*, 2019; Simpson *et al.*, 2015) such as negative behaviors like aggression (Overall *et al.*, 2016) as well as prosocial behaviors (Körner, Schütz, *et al.*, 2022). We focus on felt power because the perception of being able to exert influence has been reported as more predictive of relationship quality (RQ) than objective power indices per se (Körner & Schütz, 2021; Weisfeld *et al.*, 1992).

Most major power theories do not directly predict how power balance relates to RQ but, for example, propose that relatively less power within a relationship is linked to inhibition, feelings of constraint, and negative experiences (Keltner *et al.*, 2003; Magee & Smith, 2013; Simpson *et al.*, 2015), which is why negative implications may be assumed for the lower power partner. Only Dyadic Power Theory (Dunbar *et al.*, 2016) explicitly predicts that perceived power balance is positively associated with RQ, an assumption that has received some empirical support and has important implications for practice (e.g., couples therapy).

The link between power and RQ has also been intensively studied in the interpersonal relationship literature (e.g., Conroy *et al.*, 2016; Gray-Little & Burks, 1983). In this literature RQ is usually understood as a broad construct comprising several dimensions such as trust, admiration of the partner, and long-term potential of the relationship with a higher-order factor of global RQ (Fletcher *et al.*, 2000). Several researchers have reported that power balance is positively related to RQ (Bentley *et al.*, 2007; Blood & Wolfe, 1960; Brezsnyak & Whisman, 2004; Caldwell & Peplau, 1984; Conroy *et al.*, 2016; Drigotas *et al.*, 1999; Gray-Little & Burks, 1983; Gray-Little *et al.*, 1996; Lindová *et al.*, 2020; Oyamot *et al.*, 2010; Steil, 1997; Weisfeld *et al.*, 1992; Whisman & Jacobson, 1990) whereas only a few studies report nonsignificant links (Körner & Schütz, 2021; Sprecher & Felmlee, 1997). Power balance in romantic relationships has also been linked to reciprocity and commitment, the absence of coercive power strategies, and thus less exploitation and objectification (Gray-Little & Burks, 1983). In fact, the idea of egalitarian partners matches modern relationship ideals. Therefore, power balance was suggested as a pathway to RQ (Knudson-Martin, 2013).

Power dynamics play a critical role in how dyads function and evolve:

Equal Power: Seen in symmetrical dyads like friendships, where decisions and responsibilities are shared. Example: Two colleagues collaborating on a project as equals. : Equal power encourages fairness, collaboration, and equality in relationships

Unequal Power: Occurs in complementary dyads, where one individual has more authority or influence. Example: A parent-child relationship, where the parent has decision-making power but guides the child for their benefit. Unequal power, when used constructively, can provide structure and guidance. However, misuse of power can lead to conflict and dissatisfaction.

Dyad in social relations

In sociology, a **dyad** is a group of two people, the smallest possible social group. As an adjective, "dyadic" describes their interaction. The pair of individuals in a dyad can be linked via romantic interest, family relation, interests, work, partners in crime, and so on. The relation can be based on equality, but may be based on an asymmetrical or hierarchical relationship (master–servant). The strength of the relationship is evaluated on the basis of time the individuals spend together, as well as on the emotional intensity of their relationship. The term dyad is from Ancient Greek δυάς *(duás)* 'pair'.

A dyad can be unstable because both persons must cooperate to make it work. If one of the two fails to complete their duties, the group would fall apart. Because of the significance of marriages in society, their stability is very important. For this reason marital dyads are often enforced through legal, economic, and religious laws.

Dyads and groups occupied the theoretical and empirical attention of social psychologists in the early and mid-20th century (e.g., Becker & Useem, 1942; Murphy, Murphy & Newcomb,1937; Simmel, 1955), but as the cognitive revolution ensued (Miller, Galanter, & Pribram, 1960;Neisser, 1967) interest was redirected to social cognition. While that revolution freed psychology from a myopic focus on connections between observable stimuli and responses, many social psychologists abandoned the textured interactions of people, and directed attention to social information processing. The theoretical and empirical yield of that revolution has been substantial (Fiske & Taylor, 1991), and paved the way for a recent focus on social neuroscience(Harmon-Jones & Inzlicht, 2016). Coupled with a waning interest in dyads, was the inherent complexity of methods for dyadic research that dampened the enthusiasm of many investigators(Cronbach, 1955). The net effect was that topics once central to social psychology occupied small tranquil eddies on the edge of the social cognitive mainstream

The Nature of Dyads One of the first comprehensive analyses of dyads and their specific functions was provided by Becker and Useem (1942). Dyads were organized in two superordinate categories:"comprehensive dyads" and "segmentalized dyads." Within each were dyads that serve different social functions. Comprehensive dyads included friendships, sexual, and inter-generational pairs; whereas segmentalized dyads included aider-aided (e.g., therapist-client), teacher-pupil, status pairs (e.g., employer-employee), common interest (e.g. hobbyists), and patterned contact(e.g., two commuters). Respectively, they represent dyads with emotional and formal relationship.

In each, the function of the dyad varies and members often occupy different roles. A basic question is: when do two independent people become a dyadic unit? Becker and Useem (1942) offered a stringent criterion and stated "Two persons may be classified as a dyad when intimate, face to face relations between them have persisted ... for the establishment of a discernable pattern ...". In the early 21st century these criteria should be relaxed with the recognition that dyadic interactions need not be face to face. It is common for two people to meet remotely on the internet and establish "discernable patterns" of coordinated behavior and statistical dependence. When two commuters occupy the same location at the same time to accomplish their trek to a downtown location and chat about the weather, a dyad exists, and intimacy is not required. Similarly, two commuters on the platform that recognize one another, occupy the same space with the same function, but never speak are familiar strangers (Milgram, Sabini & Silver, 1992). When one of the commuters furtively avoids making eye contact with the familiar stranger there is behavioral dependence. When two people behave in a coordinated manner, rather than independently, and this produces a statistical dependence in their behavior, a psychological dyad has formed. **An implicit societal assumption is that dyadic interactions should be harmonious**. The mentor and student, marital partners, shopper and clerk each presume that the interaction should produce positive outcomes that range from pleasant to joyous. When the outcome is negative for one or both members, harmony is replaced by dislike or antipathy. In a non-intimate relationship, the diner may be pleased by the server's attentiveness, or annoyed after paying a plumber to fix a water leak that continues dripping. In intimate relationships, one divorced couple may pleasantly cooperate for child rearing, whereas another may evolve into an "antipair" (Becker & Useem, 1942) with reciprocal loathing.

Dyads within Groups : The dyad is phenomenologically unique compared to larger collections of people. In a group of two, there is less diffusion of responsibility compared to larger groups because one's outputs are readily discernable (Williams, Harkins, & Latané, 1981). Dyads offer the possibility for greater intimacy than larger groups, but also afford the possibility of greater conflict because"... circumscriptions effective in larger and more public groupings are at a minimum in a dyad"(Becker & Useem, 1942, p. 15). Differences in acceptable behavior constitute a fundamental distinction between a dyad and a group; ask the members of any romantic dyad. Cooley (1902) claimed that in intimate dyads, the two selves become intertwined with common purpose, goals and identities. This principle is well established in contemporary social psychology (Aron &Nardone, 2012). When intimate dyadic relationships end or are reconfigured, there is a fundamental alteration of the members' selves that can also impact other dyads within the group. Dyadic relationships are always changing psychologically because of circumstances or members' altered views of one another, and eventually all will end because of structural change (e.g., graduation, divorce) or death.

In marital dyads when one member of the couple dies, the other member has a higher probability of death compared with those who have not lost a spouse (Martikainen, & Valkonen, 1996). When reorganization is due to divorce, friends tend to side with one member and their relationship with the other member often ends (Greif, & Deal, 2012). Because dyadic relationships are nested within broader group structures; it is profitable to know what processes are consistent and inconsistent across them. For example, I wondered if the level of interpersonal attraction to a person is consistent across different social groups (family members, friends and co-workers; Malloy, 2018), the answer is no. Another study (Malloy *et al.*, 2004) considered whether trait judgments of a person in multiple dyads within one group (i.e., the family) are consistent across groups (i.e., friends) in collectivist cultures (also see Albright *et al.*, 1997; Church *et al.*, 2014). Again, the answer is no. These and other findings (e.g., Malloy *et al.*, 1997) suggest that some aspects of dyadic relationships are context specific. Overall, there has been insufficient attention to the consistency or inconsistency of dyadic behavior in different group contexts. Based on the limited data available, it appears that many dyadic processes are context (i.e., group) specific and is termed the social context hypothesis.

Dyadic interpersonal attraction (IA) was studied within groups of very highly acquainted family members, friends and co-workers. IA was determined by the perceiver (i.e., the heart of the beholder), the target (i.e., the heart of the beheld), and in specific dyads, by the unique combination of the two. The consistency of one's attraction to others and others' attraction to the person across groups were addressed using the key person design. Attraction to a person in one group was independent of attraction to that person in another, although people predicted that members of different groups were similarly attracted to them (Melloy, 2017).

Friendship maintenance behaviors of positivity, supportiveness, openness, and interaction were identified as key factors. Furthermore, maintenance behaviors differed by friendship status (best, close, or casual friends) and gender. Best friends engaged in more maintenance behaviors than close or casual friends. Gender differences in maintenance behaviors were found, such that female same-sex friendships reported using more supportiveness, openness, and interaction than male same-sex friendships. The interdependence of maintenance behaviors among friendship dyads indicated that friendships are equitable, display self-other agreement, and perceive equity of maintenance behaviors. Furthermore, both individuals' self-reported behaviors predicted dyad level satisfaction and commitment.

Fundamentals of Dyadic Relationships

Person Perception

A body of research on consensus at zero acquaintance (Ambady *et al.* 1995) consistently shows that people are quite accurate in their initial impressions of others (Ambady & Rosenthal 1992). Encoding of nonverbal cues, such as facial expressions, appears to be critical in the formation of initial impressions of qualities in people that are salient for investment in time and effort required to form lasting dyadic relationships with the other person. Indeed, it has been observed that, "people rapidly make attributions from faces, such as whether a person is trustworthy" (Todorov *et al.* 2015, p. 520).

Attraction

It is through person perception that one's level of attraction to an individual is determined. First impressions may be dominated by nonverbal cues (Todorov *et al.* 2015), including facial expressions, body movements, physical characteristics, and appearance (Eastwick *et al.* 2014). Upon further interaction, assessment of verbal behavior enables an assessment of personality, attitudes, and values. An established finding from social psychology is that people tend to be attracted to individuals who express similar attitudes and values (Byrne 1971). Interestingly, perceptions of similarity have been shown to be more important for outcome behaviors than is demographic similarity (Liden *et al.* 1993, Turban & Jones 1988). Similarity in some personality types, such as positive affectivity (Bauer & Green 1996) and proactive personality (Zhang *et al.* 2012), have also been shown to enhance attraction. When similarity is perceived, liking is enhanced, which further contributes to attraction. Admiration and respect have also been identified as important determinants of attraction.

Relationship Development

Relationships form the fabric of life as we know it. Humans are and have been social beings throughout their time on earth. Not surprisingly, a voluminous body of research on relationships conducted within many diverse fields of study, including psychology, sociology, and anthropology, has emerged (Ferris *et al.* 2009). Although differences exist across types of relationships, such as romantic, family, kinship, and work, communalities exist as well. For example, respect, trust, and shared attitudes and values are salient in the development of virtually any type of dyadic relationship. One central part of relationship development is role taking and role making that occurs for both parties of a dyadic relationship (Katz & Kahn 1966, 1978). Graen (1976) relied on this work to propose the influence of formal leaders in providing assistance to newcomers in the organization.

Role making and role taking. Expanding on the work of Kahn *et al.* (1964), Dienesch & Liden (1986) and Graen & Scandura (1987) proposed models of dyadic relationship development. Initial interactions are depicted in Graen & Scandura (1987) as commencing with sent roles from the higher status member, which are either received or rejected by the lower status member of the dyad. If the sent role is accepted by the member, the member behaves accordingly and the behavior is evaluated by the superior. This process is repeated until the superior offers a working relationship to the member, which is referred to as role making. Over time, role routinization occurs and the dyadic relationship is established. An important point often overlooked in research on relationship development, especially those in which one partner is of higher status than the other, is that both parties are involved in the development of the relationship, such that a lower status individual may be proactive in contributing to the nature of the relationship that evolves.

Intention attributions. Dienesch & Liden's (1986) model of a dyadic relationship also includes an initial sent role phase (referred to as delegation) but provides more emphasis on the member's influence in the developmental process. Of critical importance to the model are the attributions made concerning their dyadic partners' behaviors during the role development process. Dienesch & Liden (1986) reasoned that high quality relationships form when both members of the dyad attribute their counterpart's behaviors as being genuine in their intent to benefit the relationship. Attributions tend to be positive when resources and support are freely exchanged between a leader and follower

Relational leadership. Brower *et al.* (2000) proposed an integration of trust and leader-member exchange (LMX) literatures in an attempt to better understand the development and maintenance of dyadic relationships; they labeled this integration of trust and LMX relational leadership. The approach stresses the importance of trust in dyadic relationships. Interestingly, trust has been identified as a key element of LMX by Graen & Scandura (1987), and trust surfaced in critical incident interviews conducted by Liden & Maslyn (1998) in the initial stages of their scale development for a multidimensional measure of LMX (also see Liden *et al.* 2015).

Maintenance and Termination of Dyadic Relationships

Psychological contract breach. Psychological contracts were defined by Schein (1970) as the expectations held by employees and employers concerning what each owes the other. Although both parties can be responsible for breach, breach generally refers to employee perceptions of promises made by the employer that have not been fulfilled.

Shocks. As Lee & Mitchell (1994) describe, shocks are dramatic events that have an immediate negative impact on employees, because they typically involve a loss of trust in a dyadic partner (Burton *et al.* 2010). A breach in a psychological contract represents a shock that may have an immediate negative impact on the quality of an interpersonal relationship.

Social Comparison

Organizational psychology research, on which our review of dyadic relationship is centered, acknowledges that social comparison is a principal element of organizational life. Festinger (1954) proposed that all individuals strive to compare themselves with similar others to evaluate their opinions, abilities, and standing. Even though the classic social comparison theory (Festinger 1954) focused mainly on an individual's cognition of social comparison, a similar notion of comparison exists at the dyadic level within organizations (see also Vidyarthi *et al.* 2010). In the realm of dyadic relationships, social comparison of one dyad with other dyads is both innate and ubiquitous, because dyads lack objective criteria to evaluate the quality of their relationship. Also, organizational dyads never exist in a vacuum as dyadic relationships are always embedded in a network of other social relationships that one or the other party maintains with the other actors in the organization (Balkundi & Kilduff 2005, Sparrowe & Liden 1997).

There also exist other independent dyads in the organization, which are of interest to the focal dyad (Eby *et al.* 2013, Ferrin *et al.* 2006). However organizational psychology research, barring a few notable exceptions, has treated and studied dyadic relationships in isolation.

Triads

- Most of us think that two is the most stable number in relationships, but actually every dyad needs a third to stabilize it. Triangulation is natural and unavoidable in every relationship, usually with children, friends, family, and even your therapist.
- Triangulation creates a "lose-lose" dynamic that prevents all three members from raising differentiation and deepening their relationships.

Triangulation is a natural and unavoidable dynamic that occurs in every intimate relationship, especially during tense periods. Triangulation occurs when both partners are not sufficiently differentiated. Differentiation is my ability to be myself and also to be an intimate relationship . When anxiety levels are beyond the capacity of the partners to contain, a third party enters the dyad (whether on their own initiation or baited by one of the partners). Triangulation can happen with a child (or children), a best friend, a sibling, a mother (or a relative), even a therapist.

Triangulation is indeed natural and unavoidable. Yet the problems begin when the triangulation is fixed and rigid. Children get triangulated all the time (especially in divorced families). When the mother and father can't get along, they start to communicate (or attack) through their child. This forces in effect the child to choose and pick sides. Sometimes, when triangulation is rigid and loaded, the child might actually develop a symptom, which will force both parents to work together "against" them.

In couple therapy, the therapist is the third. Every couple, sooner or later, will try to triangulate the therapist to join their coalition in order to convince, change, or judge their partner. So even though in the short run, triangulation lowers anxiety and stabilizes the dyad, in the long run, it prevents growth.

A patient's race and zip code influence health care delivery more than any other variables. Mental health care disparities faced by members of racial-ethnic minority groups, Medlock *et al.* contend, include "inequities in access, symptom severity, diagnosis, and treatment," as well as in treatment outcome. Removing the socioeconomic barriers to access does not level the health care playing field. Data on the significant effects of racial bias in mental health care delivery are overwhelming, too wide-ranging and copious to enumerate.

3.2 Race and ethnicity and human development. Socio-cultural and sub-cultural differences in child's environment and their impact on childrearing practices and development. Socio-cultural imbalances and their impact on development

The effects of racism permeate all levels of health care system, regardless of the arena of medicine or psychiatry being practiced Racism manifests in many forms, from individual discriminatory acts, to structural barriers, to beliefs based on unconscious ideas. We in the helping professions wish we were exempt. As mental health clinicians, we aspire to offer our patients, through treatment, an opportunity to achieve fuller and healthier lives—as Freud said, at work and play—but our own unconscious scripts guide us too.

The stress of racial discrimination exacts a biological toll on Black and brown bodies, thereby adversely affecting overall health, argues Yale historian Carolyn Roberts. Experiencing discrimination is associated with stress and self-reports of ill health. Accordingly, if patients from minority groups are matched with providers who do not acknowledge the realities of discrimination nor the challenges of living in a racialized society, the quality of care suffers . Studies document higher patient dropout rates for interracial therapy dyads . Clinician education; cultural sensitivity; and understanding of "racial or ethnic identity, cultural values, cultural mistrust, [and] therapist cultural competence and worldview. may moderate the [potentially limiting] impact of racial differences".

Socio-cultural differences in a child's environment and its effect on child-rearing practices and development.

Every culture is characterized, and distinguished from other cultures, by deep-rooted and widely acknowledged ideas about how one needs to feel, think, and act as a functioning member of the culture. Cultural groups thus embody particular characteristics that are deemed essential or advantageous to their members. These beliefs and behaviours tend to persist over time and constitute the valued competencies that are communicated to new members of the group. It also shapes parents' belief on how to take care of their offspring. Thus, having experienced unique patterns of caregiving is a principal reason that individuals in different cultures are who they are and often differ so from one another.

Child-rearing: Child-rearing practices are defined as the techniques used in a home to elicit a certain type of behaviour from a child. It is the process of promoting and supporting the physical, emotional, social and intellectual development of a child from infancy to adulthood.

Culture difference and parenting

Culture includes physical and material elements of the environment in which the child is reared and develop as well as the factors of influence that are present in the tradition, morals, customs, beliefs, norms, values, attitudes and social motives. The Culture influence child development in various ways. The cultural factors include a like family, religion, school, mass-media, press, radio, TV, movies and other organized. And the child's education in the school largely depends on the cultural cognition of the society e.g. his relationship with his peers and teachers are influence by the values of his culture.

Culture helps to construct parents and parenting, and culture is maintained and transmitted by influencing parental cognitions that in turn are thought to shape parenting practices. Parents often face a seemingly endless array of choices when it comes to child-rearing. From deciding whether or not to work to selecting breast milk vs formula, to implementing permissive or authoritative discipline, it can be difficult for parents to decide on the right course of action. Global parenting reveals that child-rearing practices in a different culture are actually quite diverse in form and cultural influence plays a profound role.

One of the most widely debated issues in parenting is whether and to what extent a child's individuality should be nurtured. Generally, there are 2 fundamental patterns in child-rearing, individualistic and collectivist, explains communication expert Marcia Carteret on dimensions of culture. Individualistic cultures emphasize self-sufficiency, while collectivist ones emphasize the dependence of individuals on the group of which they are apart. Child-rearing practices change from culture to culture, from generation to generation and from social class to social class. What is perfectly acceptable in one culture or decade, may be seen as shocking in another. American culture generally values and rears girls and boys equally, many cultures give preference to male children.

The American standard of one child/one bedroom is a recent practice; promoted in an age of smaller families and perceived affluence. The practice is not common in other cultures. Historically, in all but the wealthiest homes, children shared not just bedrooms, but beds. Even today, Hispanic immigrant families believe that young children should sleep with siblings and/or parents. The practice, called "augusto," translates literally as "being relaxed."

Practices regarding children and money vary widely. Some American parents give children money, expecting nothing in return. Other parents give an allowance, but only with established chores. Still, others don't give allowances at all but pay children for work they do for the family. American children generally use that money for whatever they want. Other cultures expect children who live at home, grown or not, to contribute to household expenses. But if that child or young adult needs financial help for something like schooling, the extended family comes to their aid, with no questions asked.

Parenting Discipline

Personal child-rearing practices are based on whether a parent believes human nature is inherently good, or inherently bad. A parent who believes his child by nature will do the wrong thing unless taught otherwise may create an authoritarian, punitive environment with strict discipline. Another parent, believing that the child's human nature is inherently good, may create the opposite extreme; an outright permissive environment where anything goes. Parents who take the middle road value their child, but create an authoritative, but the nurturing environment, where rules are enforced, but lessons are learnt by encouragement and natural consequences.

Values and attitude

Children first learn the values and attitudes practiced in their home environment, then from schools and peers. Depending upon a society's values, the education system of an entire country may support a child-rearing practice. For instance, in America, schools place more emphasis on individualism. In Japan, teachers place more emphasis on group-consciousness.

Margaret Mead studied 3 primitive societies in New Guinea, **Arapesh,** both men and women were peaceful in temperament and neither men nor women made war. Their children were also found to peaceful as parents."Among the **Mundugumor**, the opposite was true, both men and women were warlike in temperament. They used to give punishment to their children. Their children were found to be aggressive in nature, whereas, **Tchambuli** was different from both. The men 'primped' and spent their time decorating themselves while the women worked and were the practical ones. Their children were also found performing the same gender role. Culture reflects parenting and children are the reflections of their parents.

Two studies investigate the relationship between racial attitude (dis)similarity and interpersonal liking for racial minorities and Whites in same-race and cross-race pairs. In nationally representative and local samples, minorities report personally caring about racial issues more than Whites do (Pilot Study), which we theorize makes racial attitude divergence within group members especially disruptive. Both established friendships (Study 1) and face-to-face interactions among strangers (Study 2) provided evidence for the dissimilarity-repulsion hypothesis in same-race interactions for minorities but not Whites. For minorities, disagreeing with a minority partner or friend about racial attitudes decreased their positivity toward that person. Because minorities typically report caring about race more than Whites, same-race friendships involving shared racial attitudes may be particularly critical sources of social support for them, particularly in predominately White contexts. Understanding challenges that arise in same-race interactions, not just cross-race interactions, can help create environments in which same-race minority friendships flourish.

The influence of socio-cultural factors in child development:

Socio-cultural factors are the larger scale forces within societies and culture that affect the thoughts behaviour and feeling of child development of those societies and cultures. Socio-cultural factors include language, law, aesthetics (appearance), religion, values, attitudes, social organization, family, community a person's role or status among others. The home/family background: The family is the foundation from which learning activities of any child take off. The type of family environment a child is born sets the limit for life's adjustment (Isangedigi 2011). In this discussion aspect of family background are limited to family socio-economic status, parent's educational status and parental discipline, adult predominance, difference in language and the culture of community where the child comes.

Many studies the find out that family characteristic can significant affect in child development Culture of the society: Culture is very important in the society for child development. Some sub-culture regard aggressive behaviour as acts of self-defence and toughness (child militancy). Parents train their children to be tough and aggressive in order to survive in the midst of violent ethnic group. In the midst of plural marriages where there are more children than food supply, children must learn to eat at a speed that will give them an adage over the weak and slow eaters when all of them have to eat from a common fund of food supply. So cultural differences in interactions between adults and children also influence how a child behaves socially.

Religion: Religion played a significant role in creating a sense of fatalism about child development that shaped the illness and caring experience. Any mother explained that life and death is in God control. Religion has been a controversial issue in the society. And Children who are socialized within religious families and communities often have beliefs and behaviours that conflict with those of the school. Religious fundamentalists often challenge the scientific theories taught by schools about the origin of man.

Language: Language is one of the many ways through which culture affects in child development. As early as infancy, mothers from different cultures talk to their babies differently. German mothers tend to focus on their infants' needs, wishes or them as a person. Mothers of the African tribal group Nso, on the other hand, focus more on social context. This is the child interaction with other people. So, children develop receptive language abilities before their verbal or expressive language development.

Lev Vygotsky focused on the socio-cultural factors in child development which he considered the most valuable. Vygotsky believed that higher mental process in the individual have their origin in social processes. These functions are development from the cultural tool, which are got by children unconsciously, in the process of getting the cultural, historical and other information through the verbal and nonverbal interaction. According to Vygotsky, language as the most specific and important, On the one hand it is necessary mean to provide children with verbal instructions, while on the other hand it serves as an effective means for intellectual adaptation of a child. Vygotsky was sure that "learning was a necessary and universal aspect of the process of developing culturally organized, specifically human psychological function".

In the 1960s, Diana Baumrind created a typology of three parenting styles, which she labeled as authoritative, authoritarian and permissive (or indulgent). She characterized the authoritative style as an ideal balance of control and autonomy. This typology became the dominant classification of parenting styles, often with the addition of a fourth category of indifferent or neglectful parents

Contemporary child rearing and implications for human development, impact of time factor on cultural history on development

The chart below depicts the four styles of parenting and impact on children		
Parenting Style	**Brief Description**	**Outcomes for children**
Authoritative (Democratic parenting) High demandingness and responsiveness	Authoritative parents tend to: • have high expectations for achievement and maturity, but they are also warm and responsive • set rules and enforce boundaries by having open discussion, providing guidance and using reasoning • provide their kids with reasoning and explanation for their action • Allow children to have a sense of awareness and teach kids about values, morals, and goals • provide discipline that is confrontive, i.e. reasoned, negotiable, outcome-oriented, and concerned with regulating behaviors • Are affectionate and supportive • provide their children with autonomy and encourage independence • Allow bidirectional communication •Acherish their children	Based on Baumrind's research on parenting styles, children of authoritative parents tend to: • appear happy and content • be more independent • be more active • achieve higher academic success • develop good self-esteem • interact with peers using competent social skills • have better mental health — less depression, anxiety, suicide attempts, delinquency, alcohol and drug use • exhibit less violent tendencies are securely attached.

Authoritarian Parenting (Dictator parenting) High demandingness, low responsiveness	Authoritarian parents tend to: • demand blind obedience using reasons such as "because I said so" • allow one way communication through rules and orders, attempts to reason with them are seen as backtalk • use stern discipline and often employ harsh punishment, such as corporal punishment, as a way to control children's behavior • be unresponsive to their children's needs and are generally not nurturing believe in tough love	Based on Baumrind's research on parenting styles, children of authoritarian parents tend to: • have an unhappy disposition • be less independent • appear insecure • possess low self-esteem • exhibit more behavioral problems • perform worse academically • have poorer social skills •be more prone to issues with mental health • be more likely to have drug use problems • have worse coping skills • have difficulty following rules • have difficulty with self-control • have more egocentric tendencies encounter more problems in relationships and social interactions
Permissive Parenting (Indulgent parenting) Low demandingness, high responsiveness	Permissive parents tend to: • set very few rules and boundaries • they are reluctant to enforce rules be warm and indulgent but they do not like to say no or disappoint their children	Based on Baumrind's research on parenting styles, children of permissive parents tend to: • be more impulsive • have difficulty with self-regulating their emotions • encounter more delinquency and addictions problems have issues with mental health — e.g. suicidal behavior in adolescents
Neglectful Parenting (Uninvolved parenting) Low demandingness and responsiveness	Neglectful parents tend to: • not set firm boundaries or high standards • be indifferent to their children's needs and uninvolved in their lives have issues of mental health (depression, physical abuse or child neglect when they were children)	

In reviewing the styles of parenting identified in the chart above, one can see that the focus comes more from Western, individualistic society's thoughts on the role of raising children. It is important that we look at how these ideals may influence the way in which we see families who have different ideals about parenting. This is one lens to look at parenting, but we need to be more critical thinkers about the many possibilities that exist in raising healthy, productive humans in the many different societies that exist in our world.

Additional parenting styles

Overindulgent Parenting

Overindulgent parenting is providing children with too much of what "looks good, too soon, too long." Oftentimes, it appears that parents implement these strategies to fulfill their own unmet needs or feelings of neglect from their own childhood. Examples include giving children an overabundance of:

- Things or experiences that are not developmentally appropriate for the child,
- Family resources that appear to meet the child's needs but do not,
- Anything that actively harms or prevents a child from developing and achieving one's full potential, and
- Freedom with minimal boundaries and limits that are developmentally-inappropriate for the child.

Many negative outcomes have been linked to overindulgent parenting, which include, but are not limited to, children who exhibit: extreme self-centeredness, excessive degrees of a sense of entitlement, poor decision-making and coping skills, and stifled developmental growth as a result of not experiencing or learning "necessary" life lessons.

Lawnmower Parenting

Saba Harouni Lurie, a licensed marriage and family therapist defines lawnmower parenting (also referred to as "bulldozing parenting" and "snowplow parenting") simply as: "when parents remove obstacles for their kids in hopes of setting them up to be successful." Lawnmower parents have a strong desire to protect their child from any type of struggle or obstacle. As as result, they're said to "mow over" any problems their child faces, as well as prevent problems from occurring in the first place.

This may result in creating a generation that has no ideas what to do when they actually encounter a struggle. This may lead to adversely affecting children as having to face adversity, struggle, and/or failure robs them of opportunities to learn from their mistakes. Raising children who have experienced "minimal struggle" is not necessarily doing them any favors or even making them even remotely happier.

1. Authoritative (Nurturant parent model): The parent is demanding and responsive. When this style is systematically developed, it grows to fit the descriptions propagative parenting, democratic parenting, positive parenting and concerted cultivation. Authoritative parenting is characterized by a child-centered approach that holds high expectations of maturity. Authoritative parents can understand how their children are feeling and teach them how to regulate their feelings. Even with high expectations of maturity, authoritative parents are usually forgiving of any possible shortcomings. They often help their children to find appropriate outlets to solve problems. Authoritative parents encourage children to be independent but still place limits on their actions. Extensive verbal give-and-take is not refused, and parents try to be warm and nurturing toward the child. Authoritative parents are not usually as controlling as authoritarian parents, allowing the child to explore more freely, thus having them make their own decisions based upon their own reasoning. Often, authoritative parents produce children who are more independent and self-reliant. Authoritative parenting styles are mainly produced in the context of high parental responses and high demands.

Authoritative parents will set clear standards for their children, monitor the limits that they set, and also allow children to develop autonomy. They also expect mature, independent, and age-appropriate behavior of children. Punishments for misbehavior are measured and consistent, not arbitrary or violent. Often behaviors are not punished but the natural consequences of the child's actions are explored and discussed—allowing the child to see that the behavior is inappropriate and not to be repeated, rather than not repeated to merely avoid adverse consequences

Authoritarian(Strict father model) The parent is demanding but not responsive. Authoritarian **parenting** is a restrictive, punishment-heavy parenting style in which parents make their children follow their directions with little to no explanation or feedback and focus on the child's and family's perception and status. Corporal punishment, such as spanking, and yelling are a form of discipline often preferred by authoritarian parents. The goal of this style, at least when well-intentioned, is to teach the child to behave, survive, and thrive as an adult in a harsh and unforgiving society by preparing the child for negative responses such as anger and aggression that the child will face if their behavior is inappropriate. In addition, advocates of the authoritarian style often believe that the shock of aggression from someone from the outside world will affect children less because they are accustomed to both acute and chronic stress imposed by parents.

Indulgent or permissive (The parent is responsive but not demanding): Indulgent parenting, also called permissive, non-directive, lenient, libertarian,[58] or (by supporters) anti-authoritarian,[59] is characterized as having few behavioral expectations for the child. "Indulgent parenting is a style of parenting in which parents are very involved with their children but place few demands or controls on them". Parents are nurturing and accepting, and are responsive to the child's needs and wishes. Indulgent parents do not require children to regulate themselves or behave appropriately. As adults, children of indulgent parents will pay less attention to avoiding behaviors that cause aggression in others. Permissive parents try to be "friends" with their child, and do not play a parental role

Neglectful or uninvolved: The parent is not responsive and not demanding.

Neglectful parents are unaware of what their children are doing, and if they find out, they feel indifferent towards them. Sometimes parents can be neglectful because of stressors they are experiencing in their own life.

3.3 Impact of time factor and cultural history on development. Contemporary child rearing practices and their implications for human development

Contemporary Child Rearing Practices /styles

Attachment parenting: Attachment parenting is a parenting style framed by psychological attachment theory. Attachment in psychology is defined as "a lasting emotional bond between people".

Child-centered parenting: Child-centered parenting is a parenting style advocated by Blythe and David Daniel, which focuses on the real needs and the unique person-hood of each child. Research has shown that child-centered parenting is difficult to get right and has a high chance of failing, resulting in narcissistic children.

Positive parenting: Positive parenting is a parenting style which generally overlaps with authoritative parenting and is defined by consistent support and guidance throughout developmental stages

Narcissistic parenting: A narcissistic parent is a parent affected by narcissism or narcissistic personality disorder. Typically narcissistic parents are exclusively and possessively close to their children and may be especially envious of, and threatened by, their child's growing independence. The result may be what has been termed a pattern of narcissistic attachment, with the child considered to exist solely for the parent's benefit.

Nurturant parenting: Nurturant parenting is defined by characteristics of being responsive and empathetic. It is a family model where children are expected to explore their surroundings with protection from their parents.

Overparenting: Overparenting is parents who try to involve themselves in every aspect of their child's life, often attempting to solve all their problems and stifling the child's ability to act independently or solve his or her own problems.

A helicopter parent is a colloquial early 21st-century term for a parent who pays extremely close attention to his or her children's experiences and problems and attempts to sweep all obstacles out of their paths, particularly at educational institutions. Overparenting limits a child's autonomy and essential development for independence. Helicopter parents are so named because they hover overhead like a helicopter, especially from late adolescence to early adulthood, during which time developing independence and self-sufficiency is critical to future success.

Helicopter Parenting

Helicopter parenting is characterized by caregivers who are extremely over involved in their child's life due to the belief that they can protect their child's physical and/or emotional well-being. Caregivers using this approach appear overbearing and overprotective due to the close attention they pay to all of their child's problems and successes. Parents "hover overhead" by constantly overseeing or being excessively interested in every aspect of their child's life. Some contend that cell phones are "the world's longest umbilical cord" which is contributing to this phenomenon.

Many negative outcomes have been linked to helicopter parenting, which may include, but are not limited to, children and adults who exhibit:

1. stifled developmental growth as a result of not experiencing or learning "necessary" life lessons,
2. long-term mental health problems,
3. rebellious behaviors in adolescence, and
4. a lack of independence coupled with poor decision-making, motivational, and coping skills.

Affectionless control: The affectionless control parental style combines a lack of warmth and caring (low parental care) with over-control (such as parental criticism, and intrusiveness). This has been linked to children's anxiety and to dysfunctional attitudes and low self-esteem in the children, although it is not necessarily the cause.

Slow parenting: Slow parenting encourages parents to plan and organize less for their children, instead allowing them to enjoy their childhood and explore the world at their own pace. Electronics are limited, simplistic toys are utilized, and the child is allowed to develop their own interests and to grow into their own person with much family time, allowing children to make their own decisions.

Idle parenting is a specific form of slow parenting according to which children can take care of themselves most of the time, and the parents would be happier if they spent more time taking care of themselves, too.

Toxic parenting: Toxic parenting is poor parenting, with a toxic relationship between the parent and child. It results in complete disruption of the child's ability to identify themselves and reduced self-esteem, neglecting the needs of the child. Abuse is sometimes seen in this parenting style. Adults who had toxic parents are mostly unable to recognize toxic parenting behavior in themselves. Children with toxic and/or abusive parents often grow up with psychological and behavioral damage

Pathogenic parenting: It refers to parenting style practices that are so aberrant and distorted that they produce significant psychopathology in the child. This may lead to child psychological abuse (DSM-5 V995.51).

Dolphin parenting: Dolphin parenting is a term used by psychiatrist Shimi Kang and happiness researcher Shawn Achor to represent a parenting style seen as similar to the nature of dolphins, being "playful, social and intelligent".

Traditional Parenting Style: The traditional parenting style is more commonly used in families with non-Western cultural values. Parents using this approach expect their children to respect and obey authority (e.g., parents, elders, etc.) and comply with their cultural beliefs and values without questions. Parents using this approach are high in demanding, warmth, and responsiveness, similar to the authoritative approach, however, they do not engage in democratic discussions.

'Ethnic minority' parenting style: 'Ethnic Minority' parenting style is an ethnocentric term coined in the USA out of Authoritarian parenting, and it refers to a style characterized by exceptionally high academic achievements among children from Asian backgrounds. Ethnic Minority style differs from strict authoritarian parenting by being highly responsive towards children's needs, while also differing from authoritative parenting by maintaining high demands, and not placing children's needs as a priority. This style promotes high demandingness and high responsiveness together to produce high academic performance in children.

Alloparenting parenting style: Alloparenting is the practice of co-parenting a child by biological parents and members of the extended family or community. This type of parenting is most prevalent in Central African countries, such as the Democratic Republic of Congo and the Central African Republic; especially in Akka foraging communities. Alloparenting is considered to help alleviate parental burdens by utilizing the community and allowing biological parents more time to work or participate in social events.

Unconditional parenting: The unconditional parenting style is one where parents provide their children with love and support no matter what the situation. This type of parenting does not involve rewards or punishments but instead focuses on building a strong relationship with your child. It can be beneficial as it provides a sense of security for children.

Commando Parenting is another style where parents essentially do whatever it takes to raise children in their desired way.

Gentle Parenting: In The Gentle Parenting Book, Sarah Ockwell-Smith describes the discipline of gentle parenting as "being responsive to children's needs" and "recogniz[ing] that all children are individuals." With the help of sensitivity, respect, and understanding as well as the establishment of sound limits, gentle parenting aims to raise children who are self-assured autonomous, and content.

Parenting Neurodivergent Children: Baumrind recommends implementing an accountable parenting approach. Research findings indicate that ADHD children are more negative and are often outspoken and dictatorial and show fewer inclinations to fixing issues.

Socio-cultural context

Besides the family, home ,the only setting that serves as a comprehensive context for human development from the early years onward is the children's institution. From an ecological perspective, the existence of such a context is important because it provides an opportunity to investigate the impact of a contrasting primary setting on the course of development through childhood, adolescence, and sometimes beyond, into the middle years and oldage. Unfortunately for our purposes, most investigations of development in institutions have, in keeping with the characteristic focus of the traditional model, concentrated on psychological outcomes for the individual with almost no attention to the structure of the immediate environment, or in our terms, of the microsystem in which the individual is embedded. Little information is provided about the complex of activities, roles, and relations that characterizes the institutional setting and differentiates it from the more common developmental context of family and home.

The parent-child relationship does not occur in isolation. Sociocultural characteristics, including economic hardship, religion, politics, neighborhoods, schools, and social support, can also influence parenting. Parents who experience economic hardship tend to be more easily frustrated, depressed, and sad, and these emotional characteristics can affect their parenting skills. Culture can also impact parenting behaviors in fundamental ways. Although promoting the development of skills necessary to function effectively in one's community, to the best of one's abilities, is a universal goal of parenting, the specific skills necessary vary widely from culture to culture. Thus, parents have different goals for their children that partially depend on their culture.

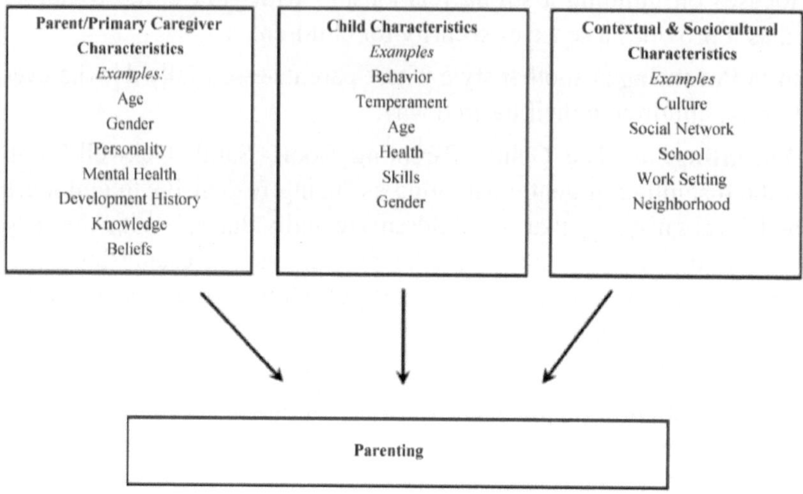

Parent-child relations in a cross-cultural perspective

A central precondition for a harmonious relationship can be seen in attachment (e.g., secure, insecure, avoidant) which also constitutes a basis for interpreting the relation between oneself and the environment ("internal working model"). Studies show that measurements of mothers' sensitivity should include culture-specific functions, e.g., proactive and reactive behavior of mothers when interacting with their child (Friedlmeier & Trommsdorff, 1999; Park *et al.*, 2012; Trommsdorff *et al.*, 2012; Trommsdorff & Friedlmeier, 1993, 2010). Several studies have also focused on the intergenerational transmission of attachment quality; however, it seems that parent-child congruence might not be the only outcome of this process as complementarity between different attachment styles of parents and children has also been found (see Lubiewska, 2013, for an overview).

Cross-cultural studies on the relations between caretakers' beliefs, childrearing, and child development in different cultural contexts conclude that the relationship context (and its culture-specific meaning) has to be taken into account in order to understand bidirectional processes between parents and children, and its effects on child development.

The role of the socio-cultural context for intergenerational value transmission

Direct context effects. Firstly of all, the socio-cultural context might have a direct influence on both parents' and their children's values. Second, the socio-cultural context might also provide similar living conditions for children and their parents that influence the value orientations of both by producing similar needs and resources. A large part of the high similarity between grandmothers and parents was due to a shared cultural stereotype or due to similar living conditions in adulthood. Evidently, needs of (grand)parents and adult children might be more similar to each other than those of adolescents and parents. Therefore, when comparing the transmission of values between generations in different stages of life, such socio-developmental aspects have to be considered as well.

Indirect context effects. Family relations are embedded in a certain socio-cultural context that provides specific conditions for the transmission process.

Cultural and Ethnic Variations In Parenting Styles

Child development is a dynamic, interactive process. Every child is unique in interacting with the world around them, and what they invoke and receive from others and the environment also shapes how they think and behave. Children growing up in different cultures receive specific inputs from their environment. For that reason, there's a vast array of cultural differences in children's beliefs and behaviour.

Language is one of the many ways through which culture affects development. Moreover, the content and focus of what people talk about in their conversations also vary across cultures. As early as infancy, mothers from different cultures talk to their babies differently. This can include the child's interactions with other people and the rules surrounding it. This early exposure affects the way children attend to themselves or to their relationship with others – forming their self image and identity. For example, in Western European and North American countries, children tend to describe themselves around their unique characteristics – such as "I am smart" or "I am good at drawing". In Asian, African, Southern European and South American countries, however, children describe themselves more often around their relationship with others and social roles. Examples of this include "I am my parents' child" or "I am a good student". Because children in different cultures differ in how they think about themselves and relate to others, they also memorise events differently.

Cultural effects of parenting: Parents in different cultures also play an important role in moulding children's behaviour and thinking patterns.

Typically, parents are the ones who prepare the children to interact with wider society. Children's interaction with their parents often acts as the archetype of how to behave around others – learning a variety of socio-cultural rules, expectations and taboos.

European-American children frequently provide long, elaborative, self-focused narratives emphasizing personal preferences and autonomy. In contrast, Korean and Chinese children's accounts are usually brief, relation-oriented, and show a great concern with authority. They often take a more passive role in the conversations.

Cultural differences in interactions between adults and children also influence how a child behaves socially. For instance, in Chinese culture, where parents assume much responsibility and authority over children, parents interact with children in a more authoritative manner and demand obedience from their children. Children growing up in such environments are more likely to comply with their parents' requests, even when they are reluctant to do so.

By contrast, Chinese immigrant children growing up in England behave more similarly to English children, who are less likely to follow parental demands if unwilling.

These studies have shown that authoritative parenting is associated with better psychosocial development and mental health across cultures. Research has suggested that authoritative parenting is more prevalent in European-American parents than in ethnic minority parents and that African-American and Asian-American parents are more authoritarian in their parenting practices than are white parents. Some researchers have suggested that authoritarian parenting may have positive effects on ethnic minority children's psychosocial adjustment and, in particular, academic achievement. In reviewing the available research, Steinberg (2001) has concluded that although African-American and Asian-American children are not as negatively affected by authoritarian parenting as are children from other ethnic groups, authoritarian parenting is not associated overall with positive adjustment. Authoritative parenting appears to confer some benefits in protecting Asian-American and black adolescents from engaging in deviant behavior and in promoting psychosocial development. However, authoritative parenting is not clearly associated with better academic achievement among ethnic minority youths.

Studies suggest that the parenting styles of European Americans differ from those of African Americans and Asian Americans. However, more research is needed to understand how culture and ethnicity interact with situational demands and the characteristics of individual children.

Concerns about the generalizability of parenting styles across diverse ethnic and cultural contexts, as well as a more general movement towards understanding the dimensions that underlie different parenting styles, has lead to a great deal of research that has focused on disaggregating Baumrind's parenting styles into their component parts. The goal of this research is to better understand how different parenting processes and behaviors interact to affect various child outcomes in different social contexts. Later theorizing and research have focused on separating parenting styles from parenting practices and differentiating forms of parental control.

Ethnic and minority parenting

It is clear that the central issues for ethnic and minority parents reflect a complex combination of group, individual and contextually derived processes. Group processes are embodied in the traditional childbearing formulations on which parents draw according to their own ethnic background. Several approaches to the study of ethnic and minority parenting represent a deviation from the deficit model that has dominated most of the field until now, shifting away from a social pathological perspective to one emphasizing the resilience and adaptiveness of families under a variety of social and economic conditions. Within these frameworks, most goals of parenting may be seen as universal, but how these goals are accomplished may vary based on context. Research on ethnic and minority families needs to be integrated into normative views of parenting in general.

Cultural effects on children

Most studies, mainly in English-speaking countries, show that children of authoritative parents have the best outcomes in different domains (behavioral, psychological and social adjustment…).The case might be different, however, for Asian populations, where the authoritarian style was found as good as the authoritative style. On the other hand, some studies have found a superiority of the indulgent style in Spain, Portugalor Brazil, but the methodology of these studies has been contested. More recently a study has shown that in Spain, while using the same questionnaire used in other countries, the authoritative style continues to be the best one for children. Furthermore, a systematic review has shown that the results do not depend on the culture but on the instruments used: studies measuring control as coercion find a detrimental effect of such control on adolescents, and better outcomes for children of permissive parents; however, when behavioral control is measured, such control is positive, and authoritative parents get the best results

Coll, C. G., & Pachter, L. M. (2002). Ethnic and minority parenting. In M. H. Bornstein (Ed.), Handbook of parenting: Social conditions and applied parenting (2nd ed., pp. 1–20). Lawrence Erlbaum Associates Publishers.

- Black, Hispanic parents are more likely than White, Asian parents to see being a parent as the most important aspect of who they are
- Black, Hispanic parents are more likely than White, Asian parents to say parenting is always rewarding, enjoyable
- White parents are the most likely to feel judged by other parents in their community
- Asian parents place more importance than other parents on college graduation, marriage for their children. Hispanic parents express higher levels of concern about their children experiencing a variety of challenges.

Alloparenting

Alloparenting, defined as care provided by individuals other than parents, is a universal behavior among humans that has shaped our evolutionary history and remains important in contemporary society. Dysfunctions in alloparenting can have serious and sometimes fatal consequences for vulnerable infants and children. In spite of the importance of alloparenting, they still have much to learn regarding the underlying neurobiological systems governing its expression.

In humans, we see a notably different system of facultative fathering, where biological fathers may or may not provide investments into their children, combined with a range of additional caregivers beyond the biological parents. These additional caregivers, or alloparents, can include siblings, grandparents and extended kin, as well as non-relatives such as step-parents, friends and neighbours.Support from alloparents, meaning "other parents," is arguably an obligate human characteristic. During childhood and adolescence, one experiences a prolonged period of physical growth and skills development, making us depended on sustained support from parents and alloparents to survive, develop and successfully reach adulthood. Non-parental caregivers are therefore necessary for successful reproduction and childrearing in humans- although who supports parents and how varies cross-culturally.

The evolution of alloparenting in humans When considering the evolution of alloparenting in humans, it is important to note that alloparenting behaviour in itself has been observed across many species. Alloparenting in humans is arguably a defining feature of human childrearing, and is observed across human populations: In the Aka foragers, infants have been found to have, on average,21 different caregivers which include kin and non-kin (Meehan, Quinlan, & Malcom, 2013). In the Hausa of Nigeria (sub-Saharan agriculturalists), children had 4.8 different caregivers on average (LeVine, *et al.*, 1996). In contemporary developed populations, parents rely on relatives such as grandparents, aunts and uncles for childcare, as well as more formal arrangements such as day-care centres and crèches (Allen, 2003; Langer & Ribarich, 2007).

Culture as a constraint to development

Geert Hofstede, a social psychologist and anthropologist poses culture as —the collective programming of the mind which distinguishes the members of one group or category of people from another‖ (Dalby, 2013). According to Iraki (2010), culture encapsulates, without being reduced to what a community values or cherishes in a particular time. It is also entails, norms and standards, which regulate the adherence to those values or desirables. It is argued that culture epitomizes our lives, desires, frustrations, ambitions and freedoms we seek. Andrews (2009) articulates that culture influences economic behaviour, political participation, value formation and evolution, and it also gives people a sense of identity or association. Fukuyama (1997) theorizes that the ability of a society to unite people depends on their norms and shared values. Where people are closer to each other in terms of ethical values, they form groups more naturally and more easily, and consequently these groups are also more likely to generate income. The evolution of culture emanates from the idea of a shared social objective (Rodney, 1973). Kaaliku (2005) explores what is meaningful to the people and what they can identify with. He attempts to: establish whether capital formation or the balance of goods are socially and culturally meaningful; whether the people identify with such issues, wealth, and whether a pacific person would rather help his family, relatives and friends. This, he called the pacific paradox and it showed that culture is an important aspect of development and will affect development and what we do. If culture is not taken into account as well as a deep understanding of the interplay between culture and development, progress remains slow. Different paths of development should be informed by recognition of how cultural factors shaped the way in which societies conceive their own future and choose the means to achieve those futures. Development embraces not only access to goods and services but also the opportunity to choose a full, satisfying, valuable and valued way of living together, the flowering of human existence in all forms and as a whole. The choice can also be for a different style of development; a different path based on assessed resources and different values from those of the highest income countries.

Recognizing Cultural Influences on Child Development

Culture influences development from the moment we're born, making an impact on us as we grow. For instance, culture can affect how children build values, language, belief systems, and an understanding of themselves as individuals and as members of society. Children can receive these cultural influences in different ways, such as through their parents, their environment, and the media.

Cultural background gives children a sense of who they are. The unique cultural influences children respond to from birth, including customs and beliefs around food, artistic expression, language, and religion, affect the way they develop emotionally, socially, physically, and linguistically.

When a child's self-identity is at odds with the social environment due to cultural differences, it can hinder learning. Fortunately, culturally competent educators help children of all cultural backgrounds learn by showing an understanding and acceptance of diverse cultures and how they make each child uniquely valuable. To understand the environment's impact on a developing child, let's look at the three main ways children process the information around them as they grow.

Classical conditioning. Drawing associations between a stimulus and response. For example, children in religious families might associate bedtime with prayers.

Operant conditioning. Drawing associations between a reward and an action. For example, children might receive dessert after eating their vegetables.

Observational learning. Absorbing and copying what they see from others in real life or in the media. For example, a child might say, "Time to clean up" because a teacher says it in school.

Children learn, therefore, by observing and making associations about their surroundings. Exposure to positive influences can favorably impact a child's development, while exposure to toxic or stressful influences can negatively impact development. Controlling the other factors , all else being equal, a child's cultural influences at birth are neutral. All too often, however, some elements of cultural background may not be accepted or understood by the society in which a child grows up — potentially harming a child's self-image and development.

In other words, the social cues a young child takes in from others about cultural background can help or hamper development because developing children readily internalize what they see and hear. When a young child's cultural background differs from the prevailing culture — for example, the child's family might speak a different language at home, eat different foods, or observe different holidays — it can affect self-image. This is especially the case if peers or even teachers treat the child in a way that reveals bias or casts the child in the role of an outsider.

According to NAEYC, childhood exposure to dominant social biases — such as favoring people who are white, Christian, heterosexual, able-bodied, thin, wealthy, fluent in English, natives rather than immigrants — can trigger developing children to judge themselves unfavorably by the same limiting standards. When children do that, their development suffers.

Parental Influences on Child Development

Parents' culture can influence their children's development. A 2019 study, for example, found that cultural values often influence the way parents raise their children, including how they discipline and set boundaries. It makes sense that parents raise their children based on cultural influences because they're preparing them to develop behaviors necessary to operate and thrive in that culture. However, when the social environment and home culture clash, developmental issues can arise.

Collectivist vs. Individualistic Cultures and Parental Discipline

Parents' cultural influences can impact how they discipline a child's behavior. This, in turn, can affect a child's development, particularly if those methods of discipline differ from the dominant cultural tradition.

There are two types of culture, a **collectivist culture** values and rewards the prioritization of community needs over individual needs, as well as generous, kind, collaborative behavior. Collectivism is the norm in Asian, Central American, South American, and African cultures.

An **individualistic culture** values and rewards assertiveness and independent action, stressing the importance of the individual over the group. Individualism dominates in North American and Western European cultures. The 2019 study cited earlier found that parents from individualistic cultures discipline differently from parents from collectivist cultures. The former group of parents might discipline their children by taking something away that matters to them personally. On the other hand, parents from collectivist cultures might tell their children to think about how their behavior affects others.

Racial identity and ethnic identity are sometimes used interchangeably. Ethnic identity development from the research literature differs from racial identity development research. The focus of racial identity is oppression and racism, whereas the focus of ethnic identity is one's shared sense of identity/ connection with others. Phinney and Ong (2007) defined ethnic identity as: "a sense of peoplehood within a group, a culture, and a particular setting" (p. 271). The psychological study of ethnic identity is informed by Erik Erikson's psychosocial stages, with identity as a process.

Jean Phinney's Three Stage Model of Ethnic Identity Development

Dr. Phinney's research about ethnic identity is shaped by Erickson's psychosocial stages, whereby identity is viewed as a process with stages. Her work is also informed by James Marcia's adolescent identity development model. Her model includes the following stages:

1. Unexamined ethnic identity: At this stage, exploring one's ethnicity might not be important to them. Familial ethnicity can be the primary influencer of one's ethnicity during this stage, and the individual at this stage may not necessarily put much thought into their own connections and shared sense of identity with particular groups, culture, or particular settings.
2. Ethnic identity search: During this stage, there is more awareness about ethnicity including exploration about one's ethnicity. During this stage this awareness about ethnicity also exposes individuals to make-meaning of their own ethnic heritage to include contrasting their familial ethnicity and learning where their own ethnic experiences fall within these experiences. As indicated by the title of this stage, they are searching for answers. Some during this stage can also experience a crisis regarding ethnicity, including being exposed to difference, oppression, racism, and other "isms."
3. Ethnic identity achievement: After searching and finding answers, the next stage is identity achievement. One feels more secure in understanding their own connections to their ethnicity; there is an increased awareness and understanding of the ethnic identities of others.

Cultural Change

- Assimilation: Change one's culture to align with the dominant culture.
- Accommodation: Partial or selective cultural change.
- Acculturation: Mutual sharing of culture.

4

Role of Systems and Policies

4.1 Children's institutions, day care and preschools as context of human development

Policies: Frameworks for decisions - should not be restrictive, should be progressive, practical and forward looking. Cultural compatibility is an important variable.

Procedures: How the policy rolls down, step by step. It has to be congruent with the organizations objectives, policies and boundary level restrictions (legal, environmental regulations).

Systems: Real life, manifestation of information flow that makes the policy live up to expectations. System also take care of inter linkages. Social change as an influencing factor of day care of children

Studies show that ECEC can contribute to child development in many ways: by improving school enrolment and retention rate and reducing school dropouts (UNICEF, 2001; World Education Forum, 2000), by narrowing the gap between children from different socioeconomic classes and, thus, 'levelling the playing field' (UNICEF, 2016, p. 41). The inequality in the development of human capabilities can and should be prevented with investments in early childhood education (Heckman, 2011).

However, for several reasons, it is not always convincing to infer about developing countries like India based on the results from high-income countries. As Kline and Walters (2016) pointed out, effect of preschool use may depend highly on a child's counterfactual activities, which heavily depends on the context. Therefore, the counterfactual use of children from developing countries is likely to differ from that of children from developed countries. Besides, preschool quality might be lower in developing countries as these countries often suffer from implementation challenges (Dean & Jayachandran, 2019).

A set of literature on the benefits of ECEC on children in developing countries shows mixed evidence (Aboud, 2006; Engle *et al.*, 2011; Hazarika & Viren, 2013; Nakajima *et al.*, 2019; Singh & Mukherjee, 2018). Research conducted so far in India on the effectiveness of ECEC provisions, particularly ICDS, can be categorised based on the effect of ECEC on the health and nutritional aspects of children (Bhasin *et al.*, 2001; Dixit *et al.*, 2018; Dutta & Ghosh, 2017; NIPCCD, 1992; Sharma & Gupta, 1993), on school-related outcomes such as enrolment and dropout. (Hazarika & Viren, 2013; Jain, 2018; Kaul *et al.*, 2015; Nandi *et al.*, 2020; NIPCCD, 2006) and on children's skill development (Dean & Jayachandran, 2019; Pandey, 1991; Singh & Mukherjee, 2018; Vikram & Chindarkar, 2020).

Delays and Deviations in Institutionalized Children's Development

Children raised in institutions often suffer from dramatic developmental delays and may follow deviant developmental pathways. However, the various causes of these delays are difficult to disentangle. First, in some instances it is difficult to know whether the institutional experience actually causes the deficits or simply maintains pre-existing deficits. Second, the forms of deprivation experienced by institutionalized children rarely occur in isolation from one another. Here we briefly present the (often severe) developmental deficiencies that most institution-reared children display. We selected to concentrate on the important areas of physical, hormonal, cognitive, and emotional development for which replicated evidence exists; however this does not constitute a complete picture. A range of other delays, deviations, and disorders have been observed in children in the post-institutional period which almost certainly had their roots in institutional care (see Chapters 5, 6, 9).

Physical Growth

Children who spend the first few years of their lives in institutional care often show retarded physical growth (Van IJzendoorn, Bakermans-Kranenburg, & Juffer, 2007; Chapter 4). Institutionalized children and adopted children with early institutional experience lag behind their family-reared peers on such central parameters of physical growth as weight, height, and head circumference. For example, in a meta-analysis of eight studies (N = 893 post-institutionalized adopted children), longer institutional stays prior to adoption were strongly and linearly associated with a more delayed age-corrected growth in height (r = .62; d = 1.71), which points to a dose-response relation, illustrating the potentially causal, negative effect of institutional care on physical growth, and in particular on height. This meta-analytic finding is supported by longitudinal studies on growth within institutions in Greece (Vorria *et al.*, 2003, see Van IJzendoorn & Juffer, 2006) and Ukraine (Dobrova-Krol, Van IJzendoorn, Bakermans-Kranenburg, Cyr, & Juffer 2008) (see also chapter 4). These effects can be long lasting. For instance, negative effects persisted to 15 years of age for adopted children who experienced extended severe global deprivation in

Romanian institutions. This was despite apparent near catch-up by age 11, which turned out later to be due, most likely, to earlier onset of the pubertal growth spurt (Sonuga-Barke, Schlotz & Rutter, 2010).

Hormonal (HPA Axis) Development (Hypothalamic-pituitary-adrenal)

Atypical patterns of diurnal cortisol activity for children living in institutions were first reported by Carlson and Earls (1997). They examined 46 children, aged 2 years, who lived in an institution in Romania. Most had been there since shortly after birth. Not one of the children exhibited a normal pattern of cortisol variation over the day (8 am, noon, 7 pm), with low early morning and slightly elevated evening values being the norm in this sample. This finding was replicated in a very small sample of post-institutionalized children, adopted from Russia and China (Gunnar, 2001). Another study conducted in a Russian Baby Home with 11 children at 3 to 5 months of age produced similar results of blunted rhythms of diurnal cortisol production (Kroupina, Gunnar, & Johnson, 1997, cited in Gunnar, 2000).

To study the effect of institutional rearing on diurnal cortisol production, Dobrova-Krol *et al.* (2008) examined 16 institution-reared children (3 to 6 years old) from Ukraine and compared them with 18 local family-reared children pair-matched on age and gender. Diurnal salivary cortisol was sampled 6 times during one day. Almost one-third of institution-reared children were chronically and severely delayed or "stunted" in their physical growth, whereas none of the family-reared children were. Institutionally-reared and family-reared children showed similar patterns of diurnal cortisol production with decreases over the day. However, non-stunted institutionally-reared children had a significantly higher total daily cortisol production than both chronically stunted institution-reared children and family-reared children. The chronically stunted children had suffered from perinatal hypoxic conditions and underwent a treatment to stabilize the functioning of the nervous system involving diazepam as well as corticosteroids (Edelstein, Bondarenko, & Bykova, n.d.), which could have had a lasting effect on the HPA-axis functioning of these children. All groups demonstrated a normal diurnal pattern with elevated morning cortisol values and subsequent decline during the day. The discrepancy with previous findings (Carlson & Earls, 1997; Kroupina *et al.*, 1997) may be explained by the older age of the children in the Ukraine study, and may be also due to the differentiation between stunted and non-stunted children.As yet, there are little or no data to evaluate whether alterations in growth or neuroendocrine activity as measured while children are in institutional care or shortly after adoption mediate any of the cognitive and emotional effects noted for post-institutionalized children (see Chapters 2, 3, 4).

Cognitive Development

The cognitive development of institutionalized children has been studied for more than 60 years. Between 1930 and 1950 a first wave of studies documented that children in institutions often showed a low IQ and severe language delays (Crissey, 1937; Durfee & Wolf, 1933), and children's orphanages have been considered "natural experiments" on the necessary conditions for intellectual growth (MacLean, 2003). Recent research continues to show the delayed cognitive performance of children in residential care (Ahmad & Mohamad, 1996; Sparling, Dragomir, Ramey, & Florescu, 2005; Vorria *et al.*, 2003; Zeanah, Smyke, Koga, & Carlson, 2005), although in one study increases in cognitive development with longer stay in the institution was noted (The St. Petersburg-USA Orphanage Research Team, 2008).

In a recent meta-analysis, children growing up in institutions showed a substantial delay in IQ compared with children reared in (foster or biological) families (Van IJzendoorn, Luijk, & Juffer, 2008). The combined effect size in 75 studies on more than 3,800 children in 19 different countries was about three-quarters of a standard deviation. For most samples, absolute IQ/DQ scores were available. The children reared in institutions showed on average an IQ/DQ of 84; the average IQ/DQ of comparison children raised in families was 104. Favorable caregiver-child ratios were associated with smaller cognitive delays, whereas early entry into residential care (before 12 months) and thus longer stays seemed to be associated with larger delays. One or more years of family life prior to institutionalization may provide a (relatively) firm basis for further intellectual development even when children grow up in a poor intellectual environment later on.

In the unique randomized control Bucharest Early Intervention Project (BEIP, Nelson *et al.*, 2007), young children living in institutions were randomly assigned to continued institutional care or to placement in foster care, and their cognitive development wastracked through 54 months of age. The authors report three main findings confirming the meta-analytic findings on non-randomized trials. First, children reared in institutions showed greatly diminished intellectual performance (borderline mentalretardation) relative to children reared in their families oforigin. Second, children randomly assigned to fostercare experienced significant gains in cognitive function. Lastly, the younger a child was when placed in fostercare, the better the cognitive outcome. Indeed, there was a continuing"cost" to children who remained in the institution for longer periods of time (see Chapter 6 on sensitive periods).

Attachment Security

Institution-reared children all experience separation from or loss of their birth parents and other caregivers. In a famous report on institutions for the World Health Organization, Bowlby (1952) concluded that children suffered from the effects of institutional care, even when their physical needs (food, clothes, etc.) were adequately met. The children are deprived of opportunities to develop stable and continuous attachment relationships due to the limited amount and poor quality of contact with their caregivers (Gunnar, Bruce, & Grotevant, 2000; Palacios & Sánchez-Sandoval, 2005; Vorria *et al.*, 2003; Zeanah *et al.*, 2005).

Six recent studies addressed the effects of institutional care on attachment, using the Strange Situation Procedure (SSP; Ainsworth *et al.*, 1978) or a modification, and institutionalized children showed high rates of insecure attachment and especially high rates of disorganized attachment (Dobrova-Krol, Van IJzendoorn, Bakermans-Kranenburg, & Juffer, 2010; The St. Petersburg-USA Orphanage Research Team, 2008; Vorria *et al.*, 2003; Zeanah, Smyke, Koga, & Carlson, 2005; Steele, Steele, Jin, Archer, & Herreros, 2009; Herreros, 2009). Overall, the attachment status distribution of institution-reared children deviated markedly from the typical distribution (62% secure, 15% avoidant, 9% resistant, and 15% disorganized; Van IJzendoorn *et al.*, 1999), with 17.1% secure, 5.5% avoidant, 4.6% resistant, and 72.8% disorganized attachments to the favorite caregiver. Indeed almost three-quarters of the institution-reared children were classified as disorganized.

The higher rate of disorganized attachment in an institutional environment of structural neglect compared to that in family-reared normative groups was to be expected. Compared to children maltreated within their families, the percentage of secure attachments is somewhat higher in the institution-reared children (maltreated children: 14%; institutionalized children: 17%) but the percentage of disorganized attachments is considerably larger (institutionalized children: 72.8 %; maltreated children: 51%, as derived from Cyr *et al.*, 2010). Whether attachment classifications, in particular disorganized attachments, mean the same in an institution as in a family environment remains to be determined (Rutter *et al.*, 2009; Zeanah *et al.*, 2005; see Chapter 3 for an extensive discussion of this issue). The wide variety of attachment outcomes within institutions, from extremely disorganized to secure, is puzzling in view of the fact that all children went through the same kind of institutional ordeal in a sensitive stage of their development. This heterogeneity is discussed in the next section. Theoretical framework

Social change has accelerated globally. Greenfield's interdisciplinary and multilevel theory of social change and human development provides a unified framework for exploring implications of these changes for cultural values, learning environments/socialization processes, and human development/ behavior. Data from societies where social change has occurred in place (US, China, and Mexico) and a community where it has occurred through international migration (Mexican immigrants in the US) elucidate these implications. Globally dominant sociodemographic trends are: rural to urban, agriculture to commerce, isolation to interconnectedness, less to more education, less to more technology, lesser to greater wealth, and larger to smaller families/households. These trends lead to both cultural losses (e.g., interdependence/collectivism, respect, tradition, contextualized thinking) and cultural gains (e.g., independence/individualism, equality, innovation, abstraction).

Social change is all-pervasive in the world, usually toward the following

- Urbanization, formal education, wealth, commerce, interconnectedness, and technology.
- These global trends shift cultural values, socialization, and development/ behavior,
- Leading to increased individualism/independence, abstraction, and innovation.
- Data from around the world reveal cultural losses as well as gains. **(Greenfield P.M., 2016)**

Model of social change, cultural evolution, and human development. Relationships for which there is empirical evidence, described in the text, have been selected for inclusion. While the horizontal arrows represent the dominant direction of social change in the world, sociodemographic change can go in the opposite direction. In that case all the horizontal arrows would be reversed.

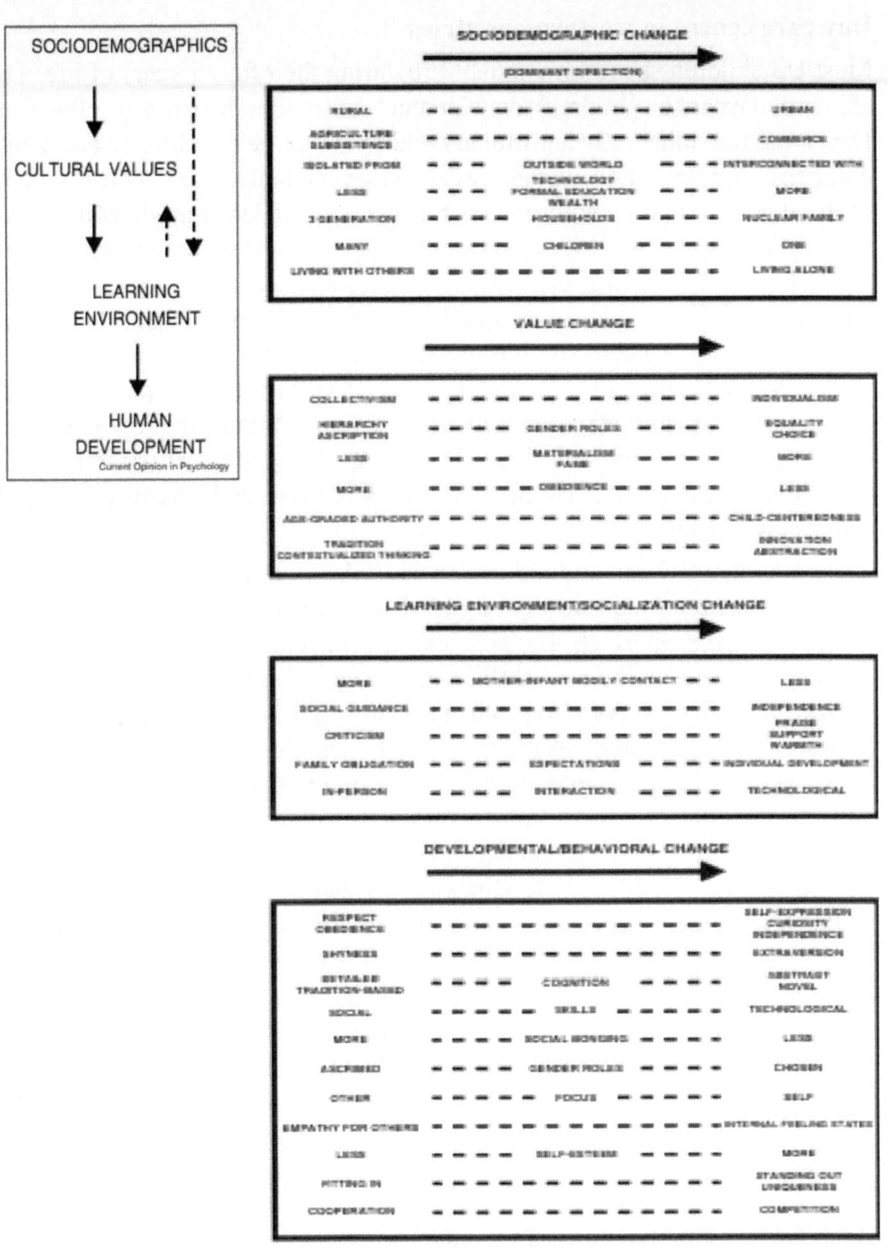

Fig. 1. A multilevel model linking sociological, cultural, environmental, and behavioral variables. Solid arrows denote the main causal pathway, with dashed arrow indicating an altemative causal pathway.

Day care centre in western countries

Most US children spend time in daycare during their first 5 years of life. The effects of daycare on children's development are smaller than family influences. Daycare is not inherently harmful to children. Daycare may help to accelerate children's cognitive and language development and buffer against the ill-effects of disadvantaged home environments. Daycare attendance predicts increased rates of infections and heightened stress, yet the developmental implications of these effects are unclear. Daycare is associated with increased child noncompliance and, when care quality is low, may also predict disruptions in parent–infant attachment. The positive effects of daycare are associated mainly with educationally stimulating, responsive care at high-quality daycare centers. High-quality daycare, however, is rare in the United States, and costs are often prohibitively high. Improving the quality, affordability, and accessibility of daycare is bound to help both children and parents thrive. Day care centre in Indian setting

Early Childhood Care and Education (ECCE) is globally recognized as a crucial element of education for all. Therefore, countries have intensely pursued it. After many efforts, progress towards ensuring ECCE is apparent in many countries in the world. But most of them could not either completely achieve this goal or compromised its quality due to inadequate resources, improper planning and ineffective implementation strategies; especially a vast and populous country like India. A number of initiatives have been taken by the Government of India for bringing quality in the form of policies, plans, constitutional amendments, acts, development of quality standards and curriculum framework, schemes and flagship programs like Integrated Child Development Services (ICDS); District Primary Education Programme (DPEP) and Sarva Shiksha Abhiyan (SSA). In this endeavour, somewhere India has compromised with the non-negotiable quality standards for ECCE, leading the emergence of expected and unexpected issues and challenges. As a result, the country has missed the target of ensuring quality in ECCE. However, recent studies give evidence of tremendous progress towards quality reforms. Of course, there are low achievements in some aspects and in some regions, but also improvement in others which shows that ensuring quality is an attainable target. This paper intends to describe initiatives of the Government of India, assess the ground realities, identify major challenges in quality reform in ECCE and suggest possible enablers to reach out the same in a study conducted by Tervahatila et al (2023) on Childcare context and socio-emotional development in toddlers – a quantitative report from the FinnBrain Birth Cohort Study, Finland revealed that, The aim of this study was to compare social competence and socio-emotional problems in two-year-old children (n=1104; girls 47.2%) who were either participating in out-of-home, center-based childcare or were cared for at home by their parents in Finland. The results showed that mothers reported more internalizing symptoms in toddlers participating in out-of-home, center-based childcare when compared to children who were cared for at home. No differences regarding externalizing symptoms or social competence were found. Overall, the results suggest that out-of-home, center-based childcare is not associated with major differences in a toddler's socio-emotional development.

National institutions and bodies for the welfare of children have been formed in practically every country in the world. Children are among the most vulnerable members of any community since they cannot vote, cannot access the influential lobbies that shape political agendas, and have limited access to the legal system and the courts, which are meant to safeguard their rights. Their requirements in terms of housing, childcare, health care, and education are crucial, and any society would incur huge expenditures if it failed to provide for its children. National Commission for Protection of Child Rights (NCPCR) and Central Adoption Resource Authority (CARA) are some of the institutions for Protection of children. Constitutional Provisions for the welfare of children- Juvenile Justice (Amendment) Act, 2015 and Child Labour (Prohibition and Regulation) Amendment Rules, 2017.

Status of Children in India

- A person who has not reached the age of 14 is considered a child according to the Child and Adolescent Labour (prohibition and regulation) Act of 1986.
- In India, children are revered as gods in their own right, yet these small deities are struggling mightily in daily social interactions.
- With the highest rate of child labour in the world, India has emerged as one of the emerging nations with the quickest rates of growth.
- In India, violence against children is pervasive and continues to be a painful reality for millions of kids from all socioeconomic backgrounds.
- In India, early marriage, domestic violence, sexual assault, domestic and school violence, trafficking, internet violence, child labour, and bullying affect both girls and boys.
- Children who experience any kind of violence, abuse, or exploitation have lifetime repercussions.
- As its name implies, the Ministry of Social Justice and Empowerment works to provide fair treatment for those segments of society who have experienced social injustices, exploitation, discrimination, and unfairness.

Institutions and Bodies for Protection of Children

Institutions and Bodies	Descriptions
National Commission for Protection of Child Rights (NCPCR)	It is a statutory organisation established by the Commissions for Protection of Child Rights (CPCR) Act of 2005 to exercise and carry out the duties entrusted to it by that Act. Its goal is to make sure that all laws, policies, programmes, and administrative procedures adhere to the child rights viewpoint as set forth in the United Nations Convention on the Rights of the Child, which India adopted in 1992, as well as the Indian Constitution.
National Institute of Public Cooperation and Child Development (NIPCCD)	A premier organisation, NIPCCD is dedicated to promoting volunteer action research, education, and documentation in the broad field of women's and children's development. Under the Societies Registration Act of 1860, it was founded in New Delhi in 1966 and operates under the direction of the Ministry of Women and Child Development. The Institute serves as the top educational facility for staff members of the Integrated Child Development Services (ICDS) programme.
Central Adoption Resource Authority	A formal creation of the Indian government is the Central Adoption Resource Development Authority (CARA). It serves as a nodal organisation for the Ministry of Women & Children and has the responsibility of monitoring and regulating domestic and international adoptions. Primarily handles adoption of orphans, abandoned children, and children who have been turned up to adoption organisations. By outlining the procedures and methods that the various adoption programme stakeholders must follow, it also provides Adoption Guidelines.
Food and Nutrition Board	The Ministry's Child Development Bureau has a technical assistance division called the Food and Nutrition Board (FNB). The nutrition-related policy matters are within the purview of FNB.It offers resources for nutrition education and awareness through training programmes and extended services.
Child Welfare Committees	The Juvenile Justice Act of 2015 recognised Child Welfare Committees (CWCs) as the ultimate district-level authority for the care, protection, treatment, development, and rehabilitation of Children in Need of Care and Protection (CNCP). As a result, CWCs are given the same powers as a first-class court magistrate and have sole jurisdiction to handle issues involving CNCP.
Child Care Institution	According to the Juvenile Justice Act of 2015, a child care institution is any Children Home, Open Shelter, Observation Home, Special Home, Place of Safety, Specialised Adoption Agency, or Fit Facility that has been approved by the Act to provide care and protection to children who require it. In Observation Homes, Special Homes, and Places of Safety, children who are in trouble with the law get residential care and protection.

CHILDLINE	Based on the 1098 Tele Helpline Model, CHILDLINE is India's first free, 24-hour emergency phone service for kids in need of help. It serves as a forum for the Union Ministry for Women and Child Development to collaborate with state governments, nongovernmental organisations, international organisations, the corporate sector, concerned individuals, and children.
POCSO e-Box for children	Protection of Children from Sexual Offences (POCSO) e-Box is an online complaint box for reporting child sexual abuse. It is an NCPCR initiative to help children report such crimes directly to the Commission.
Juvenile Justice Board	The Juvenile Justice (Care and Protection of Children) Act, 2015 ACT NO. 2 OF 2016 [31st December, 2015]
Open shelter	The open shelters referred to in sub-section (1) shall function as a community based facility for children in need of residential support, on short-term basis, with the objective of protecting them from abuse or weaning them, or keeping them, away from a life on the streets.
Foster care	The children in need of care and protection may be placed in foster care, including group foster care for their care and protection through orders of the Committee, after following the procedure as may be prescribed in this regard, in a family which does not include the child's biological or adoptive parents or in an unrelated family recognised as suitable for the purpose by the State Government, for a short or extended period of time. (2) The selection of the foster family shall be based on family's ability, intent, capacity and prior experience of taking care of children
Sponsorship.	The State Government shall make rules for the purpose of undertaking various programmes of sponsorship of children, such as individual to individual sponsorship, group sponsorship or community sponsorship.
Observation homes	The State Government shall establish and maintain in every district or a group of districts, either by itself, or through voluntary or non-governmental organisations, observation homes, which shall be registered under section 41 of this Act, for temporary reception, care and rehabilitation of any child alleged to be in conflict with law, during the pendency of any inquiry under this Act.
Special homes	1) The State Government may establish and maintain either by itself or through voluntary or non-governmental organisations, special homes, which shall be registered as such, in the manner as may be prescribed, in every district or a group of districts, as may be required for rehabilitation of those children in conflict with law who are found to have committed an offence and who are placed there by an order of the Juvenile Justice Board made under section 18.

Place of safety	1) The State Government shall set up at least one place of safety in a State registered under section 41, so as to place a person above the age of eighteen years or child in conflict with law, who is between the age of sixteen to eighteen years and is accused of or convicted for committing a heinous offence.
Children's Home	1) The State Government may establish and maintain, in every district or group of districts, either by itself or through voluntary or non-governmental organisations, Children's Homes, which shall be registered as such, for the placement of children in need of care and protection for their care, treatment, education, training, development and rehabilitation.
Adoption.	Adoption shall be resorted to for ensuring right to family for the orphan, abandoned and surrendered children, as per the provisions of this Act, the rules made there under and the adoption regulations framed by the Authority.

Conclusion

As India strives to become a superpower, its children remain far behind in terms of healthcare, education, and other amenities. Children, particularly girls, confront a lack of educational opportunities, malnutrition, infant mortality, and early marriages. Children are regarded to be kinds of deity in India, however these young Gods face major societal problems. India has become one of the world's fastest growing developing countries, with the biggest number of child labourers.

Constitutional Guarantees that are meant specifically for children include:

- Right to free and compulsory elementary education for all children in the 6-14 year age group (Article 21 A)
- Right to be protected from any hazardous employment till the age of 14 years (Article 24)
- Right to be protected from being abused and forced by economic necessity to enter occupations unsuited to their age or strength (Article 39(e))
- Right to equal opportunities and facilities to develop in a healthy manner and in conditions of freedom and dignity and guaranteed protection of childhood and youth against exploitation and against moral and material abandonment (Article 39 (f))
- Right to early childhood care and education to all children until they complete the age of six years (Article 45) Besides, Children also have rights as equal citizens of India, just as any other adult male or female:
- Right to equality (Article 14)
- Right against discrimination (Article 15)

- Right to personal liberty and due process of law (Article 21)
- Right to being protected from being trafficked and forced into bonded labour (Article 23)
- Right of minorities for protection of their interests (Article 29)
- Right of weaker sections of the people to be protected from social injustice and all forms of exploitation (Article 46)
- Right to nutrition and standard of living and improved public health (Article 47)

4.2 children in war and disasters

Children pay the highest price of humanitarian crisis. Armed conflict, natural disasters and other emergencies expose millions of girls and boys to unthinkable forms of violence, exploitation, abuse and neglect.

Many children are forced to flee their homes, some torn from their parents and caregivers along the way. In conflict, children may be injured or killed by explosive weapons and remnants of war, including during attacks on schools and hospitals. They may be recruited by **armed forces** – not only as fighters, but as scouts, cooks, porters, guards, messengers and more. Especially for girls and women, the threat of gender-based violence soars.

Through it all, children lose critical health, education and protection services. Their mental health and psychosocial needs are often neglected, with consequences that can last a lifetime.

1. **Protecting children from explosive weapons**
    - Armed conflicts are increasingly fought in populated areas children call home.
    - Explosive weapons kill and injure thousands of civilians each year, during and after armed conflict. Children account for roughly half of the casualties from explosive ordnance.
    - They are vulnerable for numerous reasons.
    - Armed conflicts are increasingly fought in populated areas, like cities, where children represent a large portion of the community. When explosive weapons are used in these areas, over 90 per cent of victims are civilians – many, children or their parents.
    - Child survivors endure staggering physical injury and psychological trauma. Some may lose their sight, hearing or limbs. Almost all will need critical and sustained psychosocial support. But explosive weapons also inflict severe harm on children indirectly.

- Especially in populated areas, explosives destroy vital infrastructure – like water pipelines, sanitation facilities, hospitals and schools – cutting children off from essential services. These losses magnify the threats children already face, by displacing them from their homes or exposing them to disease outbreaks and more. In protracted conflicts, for example, children are more likely to die from diarrhoeal disease linked to unsafe water and sanitation than from violence.
- Even after conflict, the effects of explosive weapons reverberate for years to come. Landmines and unexploded or abandoned ordnance (so-called "remnants of war") still cause death and injury across various parts of the world. Children are particularly vulnerable, attracted to remnants for their colourful appearance and unaware of how dangerous they are. Some of these weapons are familiar household objects that have been turned into explosives.
- No matter where they are used, explosive weapons endanger a child's most fundamental rights.

2. **Children recruited by armed forces or armed groups**

Thousands of boys and girls are used as soldiers, cooks, spies and more in armed conflicts around the world.

- Thousands of children are recruited and used in armed conflicts across the world. Between 2005 and 2020, more than 93,000 children were verified as recruited and used by parties to conflict, although the actual number of cases is believed to be much higher.
- Often referred to as "child soldiers," these boys and girls suffer extensive forms of exploitation and abuse that are not fully captured by that term. Warring parties use children not only as fighters, but as scouts, cooks, porters, guards, messengers and more. Many, especially girls, are also subjected to gender-based violence.
- Children become part of an armed force or group for various reasons. Some are abducted, threatened, coerced or manipulated by armed actors. Others are driven by poverty, compelled to generate income for their families. Still others associate themselves for survival or to protect their communities. No matter their involvement, the recruitment and use of children by armed forces is a grave violation of child rights and international humanitarian law.
- The recruitment and use of children by armed forces or armed groups is a grave violation of child rights and international humanitarian law.

- While living among armed actors, children experience unconscionable forms of violence. They may be required to participate in harrowing training or initiation ceremonies, to undergo hazardous labour or to engage in combat – with great risk of death, chronic injury and disability. They may also witness, suffer or be forced to take part in torture and killings. Girls, especially, can be subjected to gender-based violence.
- Warring parties also deprive children of nutrition and healthy living conditions, or subject them to substance abuse, with significant consequences for their physical and mental well-being.
- These experiences take a heavy toll on children's relationships with their families and communities.
- Children who have been recruited or used by armed actors may be viewed with suspicion, or outright rejected, by their families and communities.
- Whether or not children are accepted back into society depends on various factors, including their reason for association with armed actors, and the perceptions of their families and communities. Some children who attempt to reintegrate are viewed with suspicion or outright rejected, while others may struggle to fit in. Psychological distress can make it difficult for children to process and verbalize their experiences, especially when they fear stigma or how people will react.
- What's more, families and communities may be coping with their own challenges and trauma from conflict, and have trouble understanding or accepting children who return home. Communities need support to care for their returning children – just as the thousands of boys and girls who exit armed forces each year need it to rebuild their futures.

3. **Gender-based violence in emergencies**

   Gender-based violence reaches every corner of the globe. In emergency settings, GBV soars.

   - Gender-based violence (GBV) is the most pervasive yet least visible human rights violation in the world. It includes physical, sexual, mental or economic harm inflicted on a person because of socially ascribed power imbalances between males and females. It also includes the *threat* of violence, coercion and deprivation of liberty, whether in public or private.

- In all societies, women and girls have less power than men – over their bodies, decisions and resources. Social norms that condone men's use of violence as a form of discipline and control reinforce gender inequality and perpetuate gender-based violence. Across the globe, women and girls – especially adolescents – face the greatest risk.
- Gender-based violence takes numerous forms: Intimate partner violence, sexual violence, child marriage, female genital mutilation, trafficking for sexual exploitation, female infanticide, and 'honour' crimes are common – with intimate partner violence occurring at staggering rates in every country. Girls and women may also experience gender-based violence when they are deprived of nutrition and education.
- *Approximately one in three women and girls worldwide will experience physical or sexual violence in her lifetime.*
- Gender-based violence can occur at any point in a person's life, in times of peace or instability. But in crisis settings, threats soar.
- Armed conflict, natural disasters and humanitarian emergencies can significantly weaken a society's ability to protect women and girls from gender-based violence. Rates of intimate partner violence often increase in crisis settings. Many armed groups also use sexual violence as a tool of warfare to advance military or political aims. All the while, girls and women may be forced to trade sex for food, money and other resources they need to survive. And in some places, they are married off early or forcefully, to protect or care for their families.
- Survivors of gender-based violence suffer devastating short- and long-term consequences to their physical and mental health. Women and girls may experience severe physical injuries, unwanted pregnancies and exposure to HIV or other sexually transmitted infections. Depression, anxiety, post-traumatic stress disorder (PTSD), the limited ability to complete daily tasks, and suicidal thoughts are also common.
- *Survivors are often subjected to victim-blaming or ostracized from their families and communities.*
- What's more, many survivors are subjected to victim-blaming or ostracized from their families and communities due to social norms. This puts them at significant risk of poverty, isolation and further

violence. Some survivors are forced to marry their perpetrators. Others face retaliation for reporting their experiences or seeking support – including at the hands of family members. At its worst, gender-based violence can result in death, like in situations of 'honour killings'.

- Between 2005 and 2020, parties to conflict raped, forcibly married, sexually exploited, and committed other grave forms of sexual violence against at least 14,200 children. This sexual violence disproportionately affects girls, who were 97 per cent of cases from 2016 to 2020.

4. **Family separation during crisis**
   - *All children have the right to grow up safe with their parents, caregivers or loved ones.*
   - All children have the right to grow up with their families. Children in family environments are generally safer and better cared for, yet violent conflicts, natural disasters and other crises force thousands of girls and boys from their parents and caregivers each year.
   - Some children are separated from their families during the chaos of a humanitarian emergency. Others may be pulled away by parties involved in a violent conflict.
   - Children may be separated from their families for hours, days, months – even years. Those who receive early support in tracing their relatives are more likely to be reunified faster.
   - However they are separated, children suffer indescribable stress when split from their loved ones. Separation undermines children's mental health, physical health and development. What's more, it leaves them vulnerable to violence, exploitation and abuse.
   - Without knowing whom to trust, where to turn for help, or when they may be reunited, children can endure extreme fear and confusion. Those who face long-term separation may also miss out on education and proper medical care – essential to help them cope with distress.
5. **Mental health and psychosocial support in emergencies**
   - *Children exposed to conflict, natural disasters and other humanitarian crises can suffer severe psychological and social consequences.*
   - Millions of children around the world suffer unthinkable distress due to armed conflict, natural disasters and other humanitarian emergencies.

- They may be forced to flee their homes, some separated from their parents and caregivers along the way. During conflict, they may endure staggering violence, severe injuries and the threat of recruitment into armed groups. Many are cut off from fundamental services like clean water, health care and education, especially as towns become militarized and communities divided.
- Through it all, children often lack access to mental health and psychosocial support, with potentially devastating long-term effects.
- Anxiety, depression and other stress-related problems threaten their ability to grow up healthy and happy. Violence can take a lifelong toll on their emotional health, physical health and social development. If exposed in early childhood, the experience can even hamper a child's brain development.
- For adolescents, severe distress and exposure to traumatic events can lead to various harmful outcomes, including alcohol and drug abuse, low self-esteem, health issues, poor school performance, self-harm and suicide.
- *Children exposed to conflict face severe emotional stress that can result in lifelong mental health and psychosocial issues.*
- Failure to address mental health and psychosocial issues stalls a child's development and bars her from opportunities to participate meaningfully in society. All too often, children with mental health conditions suffer stigma, discrimination – even human rights violations. They may be excluded from their communities or exposed to further violence, including in the places they should be safest.
- Crisis situations also put parents and caregivers under mental and psychosocial duress, which can prevent them from providing the protection, stability and nurturing care their children need during and after an emergency.

Policy issues to support children under crisis

- Mental health and psychosocial support (MHPSS) is essential for improving the well-being of children affected by humanitarian emergencies. Efforts are needed to strengthen the resilience of children and their families, and to help them cope with adversity during and after a crisis. We adapt our services to local contexts, listening to what communities identify as their priorities.
- Creating nurturing environment that provide the conditions for a child's optimal development. We offer mental health and psychosocial support

to children, adolescents, caregivers, families and the wider community through a range of interventions that meet their complex needs.

Children and adolescents

- Setting up safe spaces that provide regular, structured activities to help children and adolescents develop skills to deal with crisis, solve problems, regulate their emotions, and form and maintain relationships. This could be through peer-to-peer groups, recreational activities, sports, and life-skills and vocational training.
- There is need to address harmful behaviours and norms in order to reduce bullying, abuse, neglect, exploitation and violence. When needed, we provide direct individual support or refer children to further specialized care.

Caregivers and families

- Helping parents and caregivers to develop the knowledge and skills to support children in distress. Raising awareness of the distress reactions that children may have during humanitarian crises, promote positive parenting knowledge and skills, and train parents and caregivers in supporting children with mental health conditions.
- Strengthening family and community support networks, facilitating the inclusion and participation of vulnerable families in communal activities and holding peer support groups for mothers and fathers.
- Ensuring parents and caregivers themselves are taken care of. Providing mental health and psychosocial support to teachers, and focus care for distressed parents and caregivers through initiatives like Psychological First Aid, parent education programmes, and peer support groups and to offer specialized care to parents and caregivers with mental health issues.

Communities and partners

- To work with community organizations to develop their capacity to support child and family well-being. Through activities like stigma-reduction campaigns and programmes that facilitate mental health care, s mental health and psychosocial needs to help communities take action.
- Reaching the most vulnerable children and caregivers who have experienced severely distressing events requires collaboration across specialized mental health, protection, education, social service and other actors.

- To build the capacity of health, education, social services and other sectors to provide child and family-friendly mental health and psychosocial support services. Facilitate child-friendly communication and spaces, along with other initiatives that help children and their families meaningfully shape the programmes that impact their well-being

5. Protecting children from sexual exploitation and abuse

- In areas affected by conflict, natural disasters and other emergencies, people trust aid workers to assist and protect them.
- The vast majority do so with professionalism and integrity. But some aid workers abuse their position of power through the sexual exploitation and abuse of those who depend on them, including children. These acts are unacceptable and violate criminal laws.
- **Sexual exploitation** refers to any actual or attempted abuse of a position of vulnerability, differential power, or trust, for sexual purposes, perpetrated by aid workers against the children and families they serve. **Sexual abuse** is the actual or threatened physical intrusion of a sexual nature, whether by force or under unequal or coercive conditions, perpetrated by aid workers against the children and families they serve.
- When a child is subjected to related misconduct at the hands of someone other than an aid worker, this is defined as **sexual violence**. Sexual violence against children occurs in every country, across all segments of society.
- Women and children in emergency settings face the greatest risk of sexual exploitation and abuse. High levels of need resulting from scarce resources, food shortages or economic insecurity can intensify the power imbalance in emergency settings, raising the possibility of sexual exploitation and abuse by humanitarian aid workers on whom communities depend, or by peacekeepers providing protection.

Sexual violence against children

- *Sexual violence knows no boundaries. It occurs in every country, across all parts of society.*
- Every year, millions of girls and boys around the world face sexual abuse and exploitation. Sexual violence occurs everywhere – in every country and across all segments of society. A child may be subjected to sexual abuse or exploitation at home, at school or in their community. The widespread use of digital technologies can also put children at risk.

- Most often, abuse occurs at the hands of someone a child knows and trusts.
- At least 120 million girls under the age of 20 – about 1 in 10 – have been forced to engage in sex or perform other sexual acts, although the actual figure is likely much higher. Roughly 90 per cent of adolescent girls who report forced sex say that their first perpetrator was someone they knew, usually a boyfriend or a husband.
- But many victims of sexual violence, including millions of boys, never tell anyone.
- *About 1 in 10 girls under the age of 20 have been forced to engage in sex or perform other sexual acts.*
- Although sexual violence occurs everywhere, risks surge in emergency contexts. During armed conflict, natural disasters and other humanitarian emergencies, women and children are especially vulnerable to sexual violence – including conflict-related sexual violence, intimate partner violence and trafficking for sexual exploitation – as well as other forms of gender-based violence.
- Sexual violence results in severe physical, psychological and social harm. Victims experience an increased risk of HIV and other sexually transmitted infections, pain, illness, unwanted pregnancy, social isolation and psychological trauma. Some victims may resort to risky behaviours like substance abuse to cope with trauma. And as child victims reach adulthood, sexual violence can reduce their ability to care for themselves and others.
- While sexual violence is fundamentally a crime of power, it is increasingly driven by economic motives. The internet has opened a rapidly growing global market for the production, distribution and consumption of child sexual abuse materials, such as photographs and videos. When online, children may be susceptible to sexual coercion and in-contact sexual abuse by offenders who attempt to extort them for content and financial gain.
- The harmful norms that perpetuate sexual violence take a heavy toll on families and communities too. Most children who face sexual abuse experience other kinds of violence. And as abuse and exploitation become entrenched, progress towards development and peace can stall – with consequences for entire societies.

Policy issues for protection of children during war and disasters

- Providing leadership and coordination for all actors involved in child protection.
- Strengthening child protection systems to prevent and respond to abuse and exploitation in humanitarian situations. Strengthen the social service workforce, support families and equip community-based groups to help keep children protected.
- Monitoring and reporting grave violations of children's rights in situations of armed conflict. Through the Monitoring and Reporting Mechanism. The documentation and verification of violations help to advocate for the protection of children and the delivery of emergency assistance, engaging with armed forces for children to be treated first and foremost as victims.
- Releasing and reintegrating children associated with armed forces and armed groups. We work with governments and other partners to prevent child recruitment, and to support the release and reintegration of all children who are used by armed forces.
- Preventing and responding to family separation, including by supporting family-based care for all unaccompanied and separated children, and reunifying separated children with their families quickly and safely, wherever possible.
- Addressing gender-based violence by ensuring that high-quality, holistic services are accessible to all survivors.
- Promoting mental health and psychosocial support by providing community-led services adapted to local contexts and implemented with the active participation of affected children, their caregivers and families.
- Preventing child injury and death from explosive weapons and remnants of war by leading technical guidance and programming on risk education, connecting victims to services, and advocating for ratification of and compliance with international humanitarian law.
- Protecting children from sexual exploitation and abuse by humanitarian workers by scaling up safe reporting channels and accessible response services.

Protection in Gender based violence

i) Addressing the underlying social drivers and environmental conditions that lead to a higher incidence of GBV in emergencies.

ii) Promoting the economic and social empowerment of women and girls. Economic initiatives include tuition for basic literacy and numeracy skills, advice on income-generating activities, cash transfer programmes, and schemes for credit and saving. Social empowerment activities may focus on women's leadership and participation in GBV programming, and the promotion of life skills.

Strategies to protect children from explosive weapons

i) Explosive weapons in populated areas: History shows that ending the use of certain weapons and military practices is possible. The Mine Ban Treaty and Convention on Cluster Munitions demonstrate the difference political commitment makes

ii) Landmines and explosive remnants of war: preventing child injury and death from landmines and explosive remnants of war, by leading technical guidance and programming on risk education.

iii) Assisting victims through medical care, provision of artificial limbs, mental and emotional support, and access to education – and help build our partners' capacity to protect and support children as well.

In case of family separation during crisis

i) Preventing family separation by helping families and communities identify vulnerable children before crisis strikes – and planning for reunification if the worst happens.

ii) Reuniting separated children with their families as quickly as possible once it is safe to do so – recognizing that the longer a separation lasts, the greater a child's risk of exposure to violence, abuse and exploitation.

iii) Establishing locations where lost children can be brought, identified and provided immediate care and safety.

iv) Once a child is safe, we help ensure that information about her circumstances and family is immediately gathered and documented, in order to trace family links and support the reunification process. While the family tracing process is underway, children may be living with extended family or other appropriate guardians. Occasionally, centre-based care is necessary at the height of an emergency.

v) Throughout this process, UNICEF and partners provide case management, psychosocial support and other services necessary to meet the needs of unaccompanied children.

4.3 Religion and caste systems, minority and deprived states and their effects on human development.

Policies to protect children against sexual harassment/exploitation Keeping children and adults safe from potential sexual exploitation and abuse by aid workers requires action – by training personnel, vetting partners and providing immediate response when abuse does occur.

Ensure that children and women are protected from sexual exploitation and abuse, and that victims are provided with the support and protection they need

Realization of Children's Rights Index

The Realization of Children's Rights Index (RCRI) is a grade between 0 and 10 that shows the level of realization of Children's Rights in a country. The lower the RCRI is, the lower the realization of Children's Rights in the country. The higher the RCRI is, the more Children's Rights in the country are real and respected.

Role of the RCRI

The Realization of Children's Rights Index is an indicative value allowing countries to be classified and graded according to the level of realization and respect of Children's Rights.

Calculating the RCRI

The Realization of Children's Rights Index is an indicator obtained by grouping statistical measurements, which have been given values determining its weight in the calculation.

This statistical tool created by Humanium is an estimation based upon sources judged to be reliable in their field. The Realization of Children's Rights Index is characterized by a willingness to have comprehensive view of Children's Rights.

In particular, the RCRI takes the following elements into account in its calculation :

- mortality for less than 5 years;
- life expectancy at birth;
- education;
- poverty;
- low birth weight;
- HIV;
- child labour;

- child marriage;
- female genital mutilation;
- registration of births;
- ecological impact on the future of children;
- rights and freedoms;
- the feeling of satisfaction with life;
- war and other violent situations…

Situation Level of Children's Rights

To simplify the RCRI, it can be visually illustrated using 5 colours indicating 5 situation levels of Children's Right :

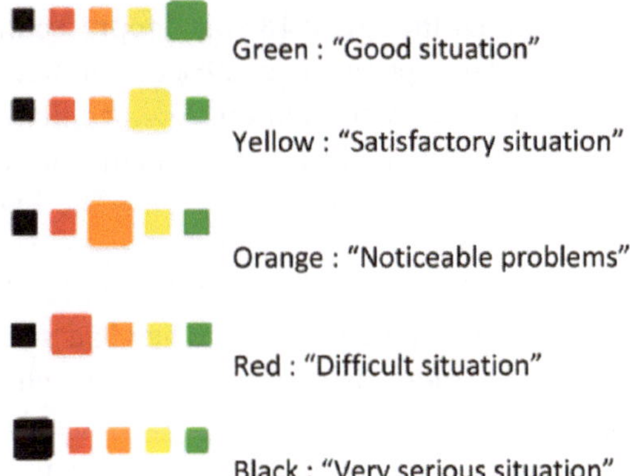

Sources

List of common sources used for country pages.

1. World Bank country data
2. UNICEF country information
3. UNICEF Situation analysis
4. Country Ministry of Health Information
5. UNDP country profiles

The Republic of India is a country in South Asia that has 29 states, it's capital state being New Delhi. It has six bordering countries including Pakistan, China, Nepal, Bhutan, Bangladesh, and Myanmar. India has a lot of diversity in customs, traditions, and languages of which Hindi is the most used. India's current population standing at 1.21 billion, it is the second most populous and seventh most extensive country in the world. It is a vast country and among the top developing nations in the world. Despite the fact that the country has shown remarkable progress in terms of economic growth, with an average of 7.3% over the past five years. It continues to face similar challenges to other BRICS countries- high growth rates being accompanied by persistent poverty and inequality. This inequality is reflected in the low human development attainments of the countries most marginalized groups including casts, tribal and rural population, women, transgendered people living with HIV and migrants. Despite India's significant progress in addressing poverty, access to education, and HIV levels, the results have mainly been uneven. India's children continue to face some of the harshest conditions anywhere in the world, with high malnutrition rates (stunting), childlabour and forced begging, and childhood illnesses such as diarrheal disease.

Status of Children's Rights

There are 472 million children in India under the age of 18 years, representing 39% of the country's total population. A large percentage, 29% of that figure constitute children between the ages of 0 to 6 years. In addition, 73% of children in India are living in rural areas, often have limited access to fundamental needs such as nutrition, access to healthcare, education, and protection. The high percentage of children living in rural areas often result in negative repressions in terms of children accessing fundamental rights. India's commission for the protection of children's rights (act 2005) (amended in 2006), has had some impact in promoting children's rights in India. Notably eliminating child labour, protection of children, and young persons. The commission's mandate is "to ensure all Laws, Policies, Programmes, and Administrative Mechanisms are in line with the Child Rights perspectives as enshrined in the constitution of India and the UN Convention on the Rights of the Child", adopted in 1989. It is clear that in India promoting children's rights is a government priority, that is enshrined within the constitution and protected in legislation. Despite this, children in India continue to face challenges in attaining these rights, particularly those related to access to education, forced labour, and child marriage. Given that children make up 39% of India's 1.21 billion population, it is imperative that the rights of these children are met.

Realization of Children's Rights Index

Indicators

An indicator is a specific observable and measurable characteristic which is used to show changes or progress of a programme is making toward achieving a specific outcome. Humanium uses key indicators to monitor the situation of children's rights globally. This is reflected in the Realization of Children's Rights Index.

Realization of Children's Rights Index :6,30 / 10
Red level : Difficult situation
Population:1.2billionPop. ages 0-14: 28.9 %
Life expectancy: 66,4 years
Under-5 mortality rate: 38 ‰

Child-sensitive Social Protection

Social protection is essential for preventing and reducing poverty for children and families, for addressing inequalities, and for realizing children's rights. In addition, it is essential that social protection programmes respond to children's vulnerabilities by optimizing positive effects for children and minimizing potential adverse consequences. Child-sensitive social protection has the opportunity to address chronic poverty, social exclusion, and external shocks which can irreversibly affect children. This is especially important for children living in rural zones which often face greater vulnerabilities exacerbated by their living conditions. Given that only 27% of Indian children live in urban zones and an overwhelming 73% live in rural areas, it is important to expand access to social protection programs for children. As a result, In India, child-sensitive social protection (CSSP) programmes are supported by saving the children, UNICEF, and the ministry of social protection. The aim is to promote and realize children's rights by ensuring that social protection measures lead to **meaningful investment in children.** Addressing the needs of Children

Right to Health

Addressing access to health is a key indicator of attaining children's rights. In India, nearly 1 million children die under the age of five, an estimated 39 deaths per 1,000 live births. Women and children are most likely to suffer disadvantages related to accessing health services such as maternal and newborn coverage. Only 1 in 3 Indian women benefit from regular monitoring of their pregnancy. In rural areas, barely 37% of births are assisted by qualified health personnel. India has more than 204 million undernourished people and Indian children remain the most affected. Children in India often face a high prevalence of stunting with rates as high as 39%. As a response, the government started a large awareness campaign in order to educate the population about the importance of a varied and balanced diet.

Children also face other challenges including a high incidence of HIV infections: 3700 new infections among children, a lack of safe drinking water, and adequate sanitation. The latter, as a result of uneven distribution of comprehensive health services to women and children in rural states.

Right to Education

Access to education in India remains a very problematic and key barrier to realizing children's rights. India continues to have the largest number of illiterate people in the world at 287 million adults, the largest population globally, and 37% of the world's total. Although India's literacy rate increased by 15% between 1991 and 2006, subsequent population growth had meant that the total number of illiterate people remained high. Despite India's efforts to devote 10.5% of its total government expenditure on education, its decentralized nature means that rich states can spend much more on education than poorer states. For example, a rich state like Kerala spent $685 per person per year on education while a poorer state like Bihar only spent $100. This unequal distribution of education further marginalizes children especially those living in rural area.

Discrimination linked to the caste system as well as discrimination against woman also remains, marginalizing millions of young Indians in the educational system. Despite this, the Indian government is trying to find solutions to allow all Indians, young or old, to benefit from high-quality education in order to fight against illiteracy. In spite of the continuing problems, India can be very proud of itself for having made considerable progress in its educational system.

Since 2009, Humanium has collaborated with local partners in India to implement children's rights, by opening residential special training centers for former child labourer's, promoting both "child-friendly villages", and various higher education aid projects. The aim of these projects is to end child labour through education, improve the living of whole villages in rural areas, and provide financial assistance projects which enable young people from disadvantaged families to pursue higher education.

Right to life

The Indian constitution of 1950 asserts that "everyone has the right to life, liberty and the security of persons", and that "no person shall be deprived of his life or personal liberty…". Despite this fundamental right enshrined in the constitution, life, survival, and child development remain areas of concern in India.

Thousands of children lose their lives each day, not only because of poverty but also because female infanticides are practiced with impunity. The main threat to Indian children's right to life stems from these female infanticides, a cultural practice that persists. In fact, each day, thousands of small Indian girls either die before being born or lose their lives because they are not desired or accepted by their family. There are several factors which contribute to the practice of female infanticides, including the dowry system which makes daughters "an unaffordable economic burden".

To deal with this problem, many Indian families turn to selective abortion of the female fetus (feticide). Even more alarming, when the birth of the child is unavoidable, families kill the babies by drowning, poison, suffocation, or deliberate negligence leading to the death of the child.

The reality is even more frightening: globally 117 million girls demographically go missing due to selective sex-abortions, and in India, every minute, 9 abortions of female fetuses will take place. Furthermore, as a result of sex-abortions India ranked number four among countries with the most skewed sex ratio at 112 males for every 100 females.

Right to protection, and freedom of expression

In India, a child has the right to be protected from neglect, exploitation, and abuse at home and elsewhere. Children have the right to be protected from the incidence of abuse, exploitation, violence, neglect, commercial sexual exploitation, trafficking, child labour, and harmful traditional practices to name a few. Yet according to a study conducted by the government in 2007, more than 69% of children aged 5 to 18 years old are victims of abuse. There are many who must face humiliation and violence every day.

More than half of the abuses inflicted upon children are committed by a close group of people who have a relationship of confidence and authority with the child. In Indian families, parents have absolute authority over their children. Furthermore, this strict discipline is also found in academic areas, a study found that 65% of school-going children have faced corporal punishment at the hands of academic staff.

A contributing factor to the neglect of children is a result of cultural values which does not have high esteem and standing for the words and opinions of children. As such, no Indian legislation specifically mentions this right, and education focuses on the respect children must show to adults.

To fully realize children right to protection, it is important to adopt a different attitude towards children and their needs. It is also necessary to invest in educating and training caregivers on children's fundamental right to protection, and prosecuting those who neglect it.

Despite the fact that the second article of the UN convention of the right of the child ensures the right not to be discriminated against including sexual orientation. Historically, in India, the LGBTQ+ community has been a target of discrimination. Mainly, as a result of a 157-year old colonial-era law (Section 377) which criminalizes certain sexual acts that are punishable by a 10-year jail term. This law not only deprived LGBTQ+ children of their fundamental rights, but subjects LGBTQ+ children and youth to bullying, harassment, isolation, and violence.

In a historical decision, India's Supreme Court has ruled that gay sex is no longer a criminal offense, overturning a 2013 judgment which upheld the law known as section 377. The court has now ruled discrimination on the basis of sexual orientation is a fundamental violation of rights. Further, the court asserted "the state had no right to control the private lives of LGBT community members and that the denial of the right to sexual orientation was the same as denying the right to privacy". In India, this ruling represents a huge victory for the LGBTQ+ community, and globally, serves as inspiration that change is possible for countries which still criminalizes homosexuality.

Right Identity

Another important factor for realizing children rights is realizing their right to identity and registration. India suffers from one of the highest non-registration rates for children in the world. Only 41% of births are registered. There is a big urban-rural difference in registration with 59% of urban children under five being registered versus only 35% in rural areas. This leads to serious difficulties for these people because they cannot benefit from child-sensitive social protection services and programmes, as such are invisible in the eyes of society.

Risk factors →Country-specific challenges

Poverty, and access to water

Since 1991, India has experienced strong economic growth. This often-spectacular development offers hints of new hopes relating to human rights and social development. However, given India's large population, many people continue to live in great poverty. The country is strongly characterized by inequalities between different regions and groups of populations. Children are most affected by poverty and social inequality. A major contributing factor to this poverty is the lack of clean water. The latter is necessary for consumption and agriculture, and it must struggle against the spread of diseases caused by the absence of sufficient sanitation. These diseases often resulting in deadly childhood illnesses.

Even though 96% of the population living in cities has access to clean water, 73% of Indian children live in rural areas where access to potable water remains a considerable problem: 20% of the rural population does not always have access to potable water. As a result of this, it is the children living in these areas who are most exposed to various health problems linked to water.

Moreover, children suffering from a lack of water miss the possibility to grow up in a healthy environment because neither homes nor schools allow them to benefit from the minimum required hygiene standards. Addressing poverty, the uneven distribution of water and other social inequalities are crucial for realizing children's rights.

Child labour

In the last years, India has put efforts in programs to fight against child labour. Major factors which contribute to this problem are the lack of food, high poverty, as well as social and economic circumstances. Other contributing factors include the lack of awareness about the harmful effects of child labour, as well as the lack of access to basic and meaningful quality education and skills training.

A recent analysis of census data in the country shows an overall decrease in child labour of only 2.2 percent yearly, over the last 10 years. Also, it has revealed that child labour has grown by more than 50 percent in urban areas.

Children under 14 often work full days in hacking cobbles stones, stitching shoes and footballs, rolling cigarettes and incense sticks, embroidery work on clothing, crafts, packing, and sticking labels to name a few. Child labour is often the result of adult unemployment or low parental wages forcing children to contribute to home production.

Children forced in labour rather than education are not given the opportunity to develop physically, intellectually, emotionally, and psychologically. India has one of the youngest populations in the world, yet more than 42.7 million children are out of school.

Child Abuse

Child sexual abuse is a dark reality that is highly prevalent in India and adversely impacts the health and wellbeing of children. Statistics show that every 15 minutes one child is sexually abused. According to research, child sex offenders can be distinguished into two groups. The first group account for about 60% of officially known offenders and show no sexual preference disorder, but who, for different reasons, sexually abuse children. The other groups are those showing a sexual preference disorder, namely pedophilia.

Individual factors such as poor socio-economic status, the death of a parent or husband, and being born to a commercial sex worker are pathways to initiation into commercial sex work. Early childhood experience because of sexual abuse was also documented as a risk factor for re-victimization as well as initiation into commercial sex work. The lack of proper family support, family and personal history of mental health pathologies, and family exposures to sexual images were some of the other potential risk factors.

Moreover, the lack of sanitation and poor safety of women were also found to be community-level factors which increased the risks for sexual abuse. The health outcomes of child sexual abuse can be grouped into mental health, physical health, behavioral, and interpersonal. Children who experience sexual abuse have a high risk for psychiatric disorders including obsessive-compulsive disorders, suicidal behaviors, and depression.

Child Marriage

In India, there was a decline in the percentages of girls getting married under the age of 16 as well as below 18 over the 20-year period 1992–2012. In addition, the mean age at marriage is 16.6 years old. There is some evidence from that child labour may in itself increase the risk of child marriage. Furthermore, girls who married as children were less likely to have been enrolled in secondary school. By the age of 15 years, only 40 percent of the girls who were child brides continued to be enrolled in school, compared to 86 percent of the girls who were unmarried when they turned 18.

4.4 impact of media, technology and social networking.

Social media is rapidly evolving in front of our eyes and it is practically difficult to reject and hide our children from this latest technology. The magnitude of media exposure is predominantly high. Survey says that and 73% of Indian children are cell phone users and the percentage of children addicted to gaming and internet has been increasing every year. In 2017, annual rate of growth of smartphone users in India is around 129%, even more than that of China (109%). Internet deaddiction centers are started in various cities of our country. Technology and social media has both positive and negative impact on children. It is the high time to comprehend the benefits and negative effects of the technology and media to utilize it effectively for the optimal growth and development of the future generation.

Social media creates the forum that enablethe users to engage in interactive web and participate, comment on and create contentas means of communicating with their social graph, other users and the public.

TYPES

Interactive

- User's inputs in turn affect the program's outputs
- Digital and electronic devices
- Software
- Multi-touch tablets

- App, video games and interactive(nonlinear)
- Screen-based media: Face book, MySpace, Gaming sites, virtual worlds,
- blogs,
- video sites such as YouTube

Noninteractive

- Media does not provide an interactive experience
- Songs, movies, and photographs
- As well as online content that cannot be modified, such as news articles
- Online videos, and banner advertisements
- Television programs, DVDs, CDs andstreaming media

Risk Factors

- Lack of self-regulation
- Lack of restrictions on creativelyexpressing oneself
- Lack of restrictions on sharinginformation
- Lack of restrictions on providingInformal support to other's work
- Generalized understanding and respectof other people's forms of expression
- Provides an alternative way of socializing.

a) Children's engagement with digital devices, screen time

1. Children's engagement with certain types of digital devices varies widely by age
2. More than one-third of parents with a child under 12 say their child began interacting with a smartphone before the age of 5
3. Nearly one-in-five parents of a child younger than 12 say their child has their own smartphone

Being able to communicate easily, getting in touch with their child are major reasons most parents say child has their own smartphone

More than a third of parents say their child under the age of 12 uses or interacts with a voice-activated assistant

Majority of parents say their child uses a voice assistant to play music; fewer use these devices to hear jokes, play games

About four-in-ten parents say they are at least somewhat concerned about data being collected about their child by voice-activated assistants

A portion of parents say their child younger than 12 uses social media; use varies by age of child, parents' level of education

Impact

Potential Harmful Effects Multi-tasking When students become addicted to thesewebsites and social media they attempt to multitask while studying. The academic performance and the ability to concentration assignment is distracted by social media addiction.

Internet addiction disorder is most commonly seen in children aged 6–19 years.As per the view of psychiatrists and counsellors, the number of people being treated for mobile addiction are mostly students in the age group of 13–24 which has shot up anywhere between 75 and 100%and above in the last one year alone and looks likely increase further in coming years. Dr Yatan Pal Singh Balhara, the psychiatrist leading the AIIMS's Behavioral Addiction Clinic says that they are planning to increase the number of days the clinic is held, given the increase in demand.

- MRI studies and CAT scans for mobile addicts have shown patterns similar to those suffering from heroin and cocaine addiction.
- National Institute of Mental Health and Neuro-Sciences (NIMHANS) clinic, Science for Healthy Use of Technology(SHUT) now gets 5 cases weekly of youngsters suffering from mobile addictive disorders compared to 1 casein every few weeks when it opened in 2014.

Time Waste : Many students are spending too much time on social sites rather than completing their homework. They may not be prepared for class. In place of studying or analyzing in depth they rush through it superficially.

Reduce in Real Life Connect

Social media is swaying people to stay online being indoor instead of going outand being social or engaging in physical activity.

E-addiction: It is the emerging problem nowadays. Magnetic resonance imaging (MRI) results reveal how addiction to social media is affecting the brain and behavior of children. Recent studies validate the reality of Internet Addiction Disorder (IAD).

Health Issues

Developmental Problems Over use of the electronic devices can lead to health issues like myopia, hearing loss, musculoskeletal problems such as body pain, back pain, carpal tunnel syndrome in children.

Eating Disorders: Many researchers have hypothesized that the media may play a central role in creating and intensifying the phenomenon of body dissatisfaction and consequently, may be partly responsible for the increase in the prevalence of eating disorders anorexia and bulimia nervosa. Prolonged use of media will result in reduced physical activity while they are interacting with the media. Food and calorie consumption is increased by the influence of food and snaking related advertisement.

Psychological Problems: Heavy television viewing was associated with self-reported symptoms of anxiety depression and posttraumatic stress. Studies show that excessive TV viewing in adolescence is a risk factor for development of depression in young adulthood. Children spend more time watching TV than spending time with their friends which leads to poor peer relationships. The risk for social isolation, anxiety disorder, agoraphobia, and antisocial behavior, including aggression and gang involvement also increases. Reportage and expose of suicide in media may increase the occurrence of suicides in children.

Neurological Problems: Survey shows that children who watch television before bed time had greater difficulty in falling asleep and had higher rate of nightmares. Heavy television viewing can trigger episodes of epilepsy in some children. Behavior Problems Children can not only observe and share feelings experienced by media characters but also respond directly to the emotionally stimulating happenings depicted in the media. Excessive viewing of television and videogames can result in behavior problem in children such as aggressive behavior, more aggressive thoughts, and more angry feelings, less empathy, fewer helping behaviors, and increase in fear. Cheating, lying, stealing, and yelling, cursing can be perceived as acceptable behavior by children.

Violence: Many violent acts are perpetrated by good guys whom kids admire so they end up in confusion to understand right and wrong. Children who watch more violence on the media are more likely to:

- Be emotionally desensitized to violence
- Avoid taking action on behalf of a victim when violence occurs
- Trust that violence is inevitable
- Have faith in violence and consider that as an acceptable way of solving conflict
- Be certain of that the world is a violent place, leading to greater anxiety, self-protective behaviors and mistrust of others
- Use violence on themselves

Exposure to Harmful Substance: Research has demonstrated a strong relationship between acquaintance to mass media messages and smoking in adolescents. Considering the potential harmful effects of mass media on attitudes and behavior, media literacy should teach youth to comprehend, evaluate, and appraise advertising and other mass media messages instead of passively remaining targets of mass media.

India challenged a lot of argument with the prohibition on on-screen smoking in films and television agendas. It has been shown that exposure to alcohol advertising and TV programming is associated with positive beliefs about alcohol consumption. Although some cross-sectional studies do not prove interconnection. A study conducted in 1990states that 56% of students in grades 5 to 12 said that alcohol advertising encourages them to drink. Findings presented that girls who had viewed more hours of TV at ages 13 and 15 sipped more wine and spirits at age 18 than those who had observed fewer hours of TV. One study has revealed independent associations between marijuana and alcohol use, and media exposure. In particular, it has stated that music exposure is associated with marijuana use while movie exposure is related to alcohol use.

Crime: The future of the current generation will be threatened by the new forms of media assisted crimes like identity theft and child pornography.

Sexual Behavior and Solicitation: Potentially damaging letters about sex in the media may central to unnatural sexual performance. Children and teenagers reflect media as the main source of data about sex which may deliver easy admittance to the sexual offenders to target children.

Cyber Bullying: Cyber bullying is a growing phenomenon in virtual communication environments Technology has given it a new platform of action at a faster speed involving a number of people at the same time. Cyber bullying can lead to great risk among the growing children enabling them in to stress anxiety, depression, low self-esteem, school dropouts and also ending their one's own life.

The Obesity Link

Kids who spend too much time using media are more likely to be overweight. Health experts have long linked too much screen time to excess weight. When they're staring at screens, kids are inactive and tend to snack. They're also see lots of ads that encourage them to eat unhealthy foods like potato chips and drink empty-calorie soft drinks that often become favorite snack foods.

Studies show that decreasing the amount of TV kids watched led to less weight gain and lower body mass index (BMI). Replacing video game time with outdoor game time is another good way to help kids maintain a healthy weight.

Watching Risky Behaviors

Characters on TV and in video games often show risky behaviors — like drinking alcohol, using drugs, and smoking cigarettes — as cool, fun, and exciting. When these things seem acceptable, kids and teens might be tempted to try them. That might lead to substance abuse problems. Media can also expose kids to high-risk sexual practices, and studies show that teens who watch lots of sexual content are more likely to start sexual activity at a young age.

Seeing Violence

The average American child will see 200,000 violent acts on TV by age 18. Many of these are done by the "good guys," whom kids are taught to admire. In fact, often the hero wins by fighting with or killing the "bad guys." This can confuse kids as they try to understand the difference between right and wrong. Kids who view violence onscreen are more likely to show aggressive behavior, and to fear that the world is scary and that something bad will happen to them.

Young kids are easily frightened by violent images. Because they don't yet understand what is real and what is make believe, simply telling them that violence isn't real won't help. Behavior problems, nightmares, and trouble sleeping may happen after kids watch violence on screen.

Older kids can be scared by violent images too. Talking with kids this age will help them, so it's important to comfort them and explain what they see to help ease fears. But it's even better to not let your kids view shows or play games that could scare them.

Potential Benefits

Better Hand-Eye Coordination: Children develops hand-eye coordination by using the application in computer and other gadgets. This skill develops when they visually follow an object on the screen and actively participate in the activity the application presents. As children begin schooling where they learn how to use their hands and eyes while writing and drawing hand-eye coordination becomes essential.

Promotes School Readiness and Cognitive Development :Technology provides opportunities for children to gain the necessary abilities to be prepare them for school. Many developmental computer programs which are available now days increase a child's readiness to learn and write. It enables them to gain competency to be successful in school and develop higher-order thinking and applied learning.

Capacity for Visual Attention Enhances: Through Gaming and learning applications, children learn to pay more attention to the details in order to complete any activities. These activities train them to pay more attention to every detail on the screen in order to achieve the goals of the game.

Motivation to Complete Tasks: Many applications and computer activities are fashioned to motivate children to complete a level in order to unlock the next level. The learning games may help to create a base for children to keep on trying and to overcome challenges even outside the screen setting.

Dynamic Spatial Skills: The dynamic spatial skills of the children are developed through interactive screen time. Through technology, they can learn the comparison of the size of cat and ant and understand the difference between man and building which enable them to understand the world around them.

Develops Horizons: Through technology Children are exposed to life, they understand beyond four walls of their home and their class room through multimedia presentations of various places and animal.

Social Skills: Human being is a social animal, keeping that in mind communication today is a boon that helps each one of us to connect to people and develop a social relationship with family, friend's, peers. This helps in a constructive positive relationship and encourages friendliness, cooperation, understanding, sympathy and acceptance from multiple groups which enhances a positive social behavior [16, 17].

Education: Educational programs conveyed through media helps the children to gain knowledge and academic abilities. Children can start learning letters of the alphabet, count or identify shapes.TV shows and movies based on books can encourage children in reading, writing and critical thinking skills. Web programs designed with educational value will engage children to develop competency in thinking, creating, problem solving, questioning, critiquing, communicating and making connections. Social media networks help youth get higher level of education, choose better career paths, and build personal relationships. Online virtual classrooms will support students to view teachers and listen and learn from lectures on YouTube, in their own time. Video games including information on academic content and problem-solving skills will augment children's learning. Development in media assisted learning and interactive game playing will lead to improvement in IQs.

Develops Skills and Talents: Social media which is interactive in nature can be used to explore children's talent. Exposing children to music and writing, will enhance the latent skills in children and serve as a guidance for parents to place or train their children as per their interest and develop skills as per their inclination. Technology presents opportunities to present different activities children may eventually engage in. For example sports, playing musical instruments learning foreign languages, etc.

Awareness: Social media can be used as a voice of reason in society. Social networking sites such as Facebook and Twitter are being used by the current young generation to create a voice for the voiceless. Certain facebook sites like "disability is not inability" give a psychological boost to disable people which encourages the society to treat disable people more human creating a positive society for disabled.

Recommendations

Educate

- Parents should have adequate knowledge about the media, technology, trends.
- Children should be educated about benefits and harmful effects of social media and help them to differentiate that.
- Educate the children about the appropriate methods to select, use, participate, and evaluate technology and interactive media in intentional and development all.
- Educate the children about the darker side of the media like cyber bullying, stalking, inappropriate materials, over sharing, etc.
- Advocate and monitor content of media
- Set aside a fixed time to use gadgets.
- Restrict phone during meal times.

Encourage

- Encourage children to spend more time in real face to face interaction which will enable them to learn more social skills, subtle vocal cues, interpret body language and adapt different personalities which are essential to survive on real world.
- Identify the child's passion and interest like sports, playing musical instruments, crafts, writing, etc. and nurture them.
- Help the child to take advantage of social networking to enhance learning, to collaborate with fellow students.

Emphasize

- Parents should emphasize to their children the difference between real-life and online relationships. For example, having 500 friends in Facebook, does not necessarily mean that he/she is popular.
- Emphasize to your child the difference between writing and spelling for social networking and real-world writing.
- Focus on the appropriateness and the quality of the content viewed.
- Provide a balance of real and online activities programs for young children.

Engage

- Parents should be engaged in the social media that their child belongs to understand the impact it can have on their child.
- Follow the child to be are aware of their activities online. Just lurk or be a silent, watchful friend. Avoid or refrain from commenting as this may turn her and their online friends off.
- Turn off mobile data and Wi-Fi before going to bed.
- Parents should update themselves to the online trends and the latest technologies, websites and applications which their children are using.
- Parents should monitor the activities of the children in the social networking sites and discuss regarding inappropriate content or connections to which appear problematic and remove them.

Express

- Express the concern of using the computer or any gargets in the presence of parents.
- Place computer in the place where parents regularly pass by.
- If the child seems upset after looking after phone or device, ask them about it.
- A study conducted by CNN states that when parents are involved with their children's social media activities, children are less likely to be troubled with what happens to them online.
- Express support for good media.

Conclusion Media and technology has both positive and negative implications on children. They cannot be eradicated however we can definitely ensure to maximize the potential benefits and minimize the harmful effects and use the resources to enhance child's growth rather than destructing their well-being

Table 1: Benefits and risks of Internet use by children and youth

Benefits of the Internet
• Near limitless information resource • High degree of availability and affordability in most communities in the western world • An asset for research or homework • A communication tool for teachers and university professors with their students • Rapid and inexpensive communication via e-mail and video linkage • Access by the disabled to much that may otherwise be unavailable: • Through technology that makes computer use possible • By taking advantage of communication with peers that does not require that their disability be visible • Source of entertainment

General risks
The Internet, by its interactive nature, is prone to use for excessive periods of time. 'Internet addiction', also known as **pathological Internet use** or **Internet addiction disorder**, is a known problem among adults and is no less a problem with children whose usage may be unlimited (30,49,50) • Instant messaging and e-mail may interfere with legitimate work (schoolwork and in the workplace) • Like television, excessive time spent on the Internet may interfere with normal socialization between child and peers, and with family • Excessive use fosters inactivity and, potentially, obesity
Misinformation • Lack of monitoring for appropriateness or accuracy of information • Health risks posed from following incorrect information • Fostering inappropriate ideas for children and teenagers such as the following: Encouraging inappropriate weight loss and diets. To read more on this, see www.eating-disorder.org/prosites.html or my.webmd.com/content/article/34/1728_85382 • Generating a sense of need to acquire material goods • False advertising or scams typically spread by e-mail • Temptation to acquire (or steal) credit cards to pay for on-line services
Adverse effects on values • Pornography: Underage children may accidentally or by design access pornographic sites • Predators have access to minors through unsupervised use of chat rooms and e-mail • Loss of inhibitions normally experienced when face-to-face, promoting sexual exchanges and promiscuity • On-line gambling • Potential for plagiarism in school work • Promotion of hatred or violence through Web sites targeting a specific group, such as women, homosexuals, and religious or ethnic groups • Use of the Internet by school bullies to ridicule or spread hatred toward a victim • Violence through video games, song lyrics and on-line bullying, access to hazardous materials or information on making weapons

Associations of media use and early childhood development: cross-sectional findings from the LIFE Child study

Schwarzer, C., Grafe, N., Hiemisch, A. *et al.* Associations of media use and early childhood development: cross-sectional findings from the LIFE Child study. *Pediatr Res* **91**, 247–253 (2022). https://doi.org/10.1038/s41390-021-01433-6

Background

Excessive media usage affects children's health. This study investigated associations between children's and mother's media use, parent–child interactions, and early-childhood development outcomes.

Methods

Two hundred and ninety-six healthy 2–5-year-old preschoolers (52.4% male, mean age = 3.5 years) and 224 mothers from the LIFE Child cohort study were analyzed. Screen times and parent–child interactions were assessed using standardized parental questionnaires. Developmental skills were investigated using the standardized development test ET 6-6-R.

Results

High screen times in children (>1 h/day) were significantly associated with lower percentile ranks in cognition ($b = -10.96$, $p<0.01$), language ($b=-12.88$, $p<0.01$), and social–emotional skills ($b=-7.80$, $p=0.05$). High screen times in mothers (>5 h/day) were significantly associated with high media use by children (OR = 3.86, $p<0.01$). Higher parent–child interaction scores were significantly associated with better body motor ($b=0.41, p=0.05$), cognition ($b=0.57$, $p<0.01$), language ($b=0.48$, $p=0.02$), and social–emotional outcomes ($b=0.80, p<0.01$) in children.

Conclusions

Public health strategies should seek to educate caregivers as competent mediators for their children's media habits, with focus on the need for children to have frequent parent–child interactions.

Impact

- High media usage in children is related to poorer cognition, language, and social–emotional skills.
- More frequent parent–child interactions are associated with better body motor, cognition, language, and social–emotional skills in children.
- High level of media use in mothers is not directly related to children's development outcomes but is directly related to high media usage of children.
- Public health strategies should seek to raise media awareness and management in both parents and children.

The study entitled "Digital media exposure and cognitive functioning in European children and adolescents of the I. Family study" is co-authored by Elida Sina, Christoph Buck, Wolfgang Ahrens, Juul M. J. Coumans, Gabriele Eiben, Annarita Formisano, Lauren Lissner, Artur Mazur, Nathalie Michels, Denes Molnar, Luis A. Moreno, Valeria Pala, Hermann Pohlabeln, Lucia Reisch, Michael Tornaritis, Toomas Veidebaum, Antje Hebestreit, and researchers associated with the I. Family consortium.

5 key findings from the study

1. More digital exposure means higher impact: The duration of exposure to digital media in children is positively associated with emotionally driven impulsiveness and cognitive inflexibility, independent of a child's well-being, weight status or sleep duration. "One additional hour of smartphone and internet exposure daily were associated with higher impulsivity score," the study says..
2. There is a difference between internet-based and non-internet-based media: The longer children used smartphones or other digital media the higher impulsivity score they received, "while no association was observed for non-internet-based media including TV and PC." The authors suggest that "perpetual flow of information and input received simultaneously from the online environment such as emails, notifications and social media posts, may act as stressors". This in turn could lead to "mindless eating in front of screens especially in reward-seeking contexts", and this may also reflect that digital media for children and adolescents displaces real-life interactions between parent and child as well as sibling and peer relationships.

   "Social interactions are crucial for a healthy development because they built the foundation of processes related with personality and cognitive development, such as emotion regulation," the research says. "The interference of digital media with parent–child interactions may compete with children's ability to concentrate and regulate their emotions, leading to internalising and externalising problems like reduced ability to control impulses."
3. High smartphone and internet use elicits different results for boys and girls: "The association between "high use of smartphone/internet use, medium TV/low PC" and impulsivity was stronger among girls compared to boys. This could be because girls mainly use smartphone/internet for socialising and social media, while boys mainly play games, the researchers say.

   "Previous evidence shows that social media exposure impacts girls' and adolescents' psycho-emotional well-being and body-image via social comparisons over images posted on these platforms," the research says. "This may lead to emotional overeating or restrained eating" owing to coping strategies for relieving negative emotions.
4. Impact can lead to poorer academic performance: Smartphone exposure and multimedia tasking were associated with higher cognitive inflexibility, which "may adversely impact youth's ability to shift

between tasks" – and this may lead to poorer academic performance in youths, say Lucia and her colleagues.

"Additionally, repeated exposure to fast-paced content, like short-edited video segments in social media (such as Instagram reels) or online game applications might trigger individuals into seeking higher arousal levels, which in turn hamper engagement in activities that require sustained attention (for example homework)."

5. Need for further research in this are: The research makes clear that it does not distinguish between the use of smartphones and Internet use for entertainment and educational purposes, and its examination of smartphone duration includes use for various content including music videos and TV shows – so the researchers urge further studies that might delve deeper into the particular use involved.

4.5 policies and programmes based on ecological factors impacting human development

Ecological-based policies and programmes in India that significantly impact human development — focusing on health, education, livelihoods, social inclusion, and ovrall well-being:

1. **National Action Plan on Climate Change (NAPCC)**

   Launched: 2008

   Focus: Sustainable development while addressing climate challenges.

   Impact on Human Development:

   - **Livelihoods:** Promotes climate-resilient agriculture (via the National Mission for Sustainable Agriculture).
   - **Health:** Reduces exposure to pollution and heat stress.
   - **Education & Awareness:** Informs communities about sustainable practices.
   - **Equity:** Helps vulnerable populations adapt to climate change (e.g., farmers, forest dwellers).

2. **Green India Mission (GIM)**

   Part of NAPCC

   Focus: Forest-based livelihoods, afforestation, ecosystem restoration.

   Impact on Human Development:

   - **Income generation** for rural communities through eco-restoration.
   - **Environmental services** (e.g., clean air, water) contribute to better health.
   - **Skill development** in eco-friendly forest-based enterprises.

3. **National Agroforestry Policy – 2014**

   Objective: Integrate trees with agricultural systems to enhance productivity and sustainability.

   Impact on Human Development:

   - Improves **food and nutritional security**.
   - Provides **supplementary income** (timber, fruits, fodder).
   - Increases**resilience to climate shocks**.

4. **Jal Shakti Abhiyan**

   Focus: Water conservation and resource management.

   Impact on Human Development:

   - **Ensures drinking water** access in rural areas.
   - **Improves health outcomes** by addressing water scarcity and contamination.
   - **Reduces water conflict**, enhancing community cohesion.

5. **Namami Gange Programme**

   Objective: Rejuvenation of the Ganga River and its tributaries.

   Impact on Human Development:

   - Improves **public health** through reduced waterborne diseases.
   - **Employment** through riverfront development and eco-tourism.
   - **Education and behavior change** via community involvement in river conservation.

6. **National Biodiversity Action Plan (NBAP)**

   Aligned with: Convention on Biological Diversity (CBD) **Focus:** Biodiversity conservation and sustainable use.

   Impact on Human Development:

   - Preserves **medicinal plants** and local biodiversity essential for health and livelihoods.
   - Promotes **traditional knowledge** and rights of indigenous communities.
   - Supports **sustainable eco-based enterprises**.

7. **Swachh Bharat Mission (SBM)**

   Objective: Sanitation, hygiene, and solid waste management.

   Impact on Human Development:

   - **Reduces disease burden** (e.g., diarrhea, respiratory infections).

- **Improves school attendance**, especially for girls.
- Promotes **community empowerment** and dignity.

8. **UJALA & PAT (Energy Efficiency Programmes)**

   Focus: LED adoption, energy-efficient practices in industries.

   Impact on Human Development:

   - **Reduces household energy costs**, freeing up resources for education and health.
   - **Improves quality of life** by extending lighting access in remote areas.
   - Encourages **behavioral change** towards sustainable consumption.

9. **Wildlife Protection & Eco-Sensitive Zones (ESZs)**

   Laws: Wildlife Protection Act, Forest Conservation Act, ESZ notifications.

   Impact on Human Development:

   - **Eco-tourism opportunities** create jobs in rural areas.
   - **Resettlement & compensation programmes** support forest-dependent communities.
   - **Environmental education** increases respect for nature and conservation.

10. **Compensatory Afforestation Fund Management & Planning Authority (CAMPA)**

    Objective: Offset deforestation by afforestation and eco-restoration.

    Impact on Human Development:

    - **Forest-based livelihoods** for tribal and local populations.
    - **Restoration of ecosystem services** (e.g., clean water, air, fertile soil).
    - **Community participation** enhances environmental stewardship.

Impacts on Human Development

Area	Ecological Contribution
Health	Clean air, water, sanitation, disease prevention (SBM, Namami Gange, NAPCC)
Education	Environmental literacy, school hygiene, awareness of sustainable practices
Livelihoods	Forest-based incomes, eco-tourism, green jobs (GIM, CAMPA, ESZ)
Nutrition	Agroforestry, clean water, climate-resilient agriculture (NMSA, Jal Shakti)
Social Equity	Empowering vulnerable communities through inclusive conservation programmes
Resilience	Climate adaptation, disaster risk reduction, sustainable farming

4.6 Research and assessment of human development based on ecological perspectives

Introduction

Human development, traditionally assessed through indicators such as health, education, and income, is increasingly being analyzed in conjunction with ecological factors. Ecological factors refer to the natural environment, resource availability, ecosystem services, and the impact of environmental sustainability on quality of life. These components play a critical role in shaping development outcomes, particularly in vulnerable and resource-dependent communities.

1. **Key Ecological Factors Influencing Human Development**
    - **Natural Resource Availability:** Access to clean water, fertile soil, forests, and biodiversity supports livelihoods, food security, and health.
    - **Climate and Weather Patterns:** Extreme weather events, temperature shifts, and climate change influence agriculture, migration, and public health.
    - **Environmental Quality:** Pollution levels, waste management, and air and water quality directly impact health outcomes.
    - **Ecosystem Services:** Benefits derived from ecosystems, such as pollination, water purification, and climate regulation, support sustainable living.
    - **Land Use and Degradation:** Deforestation, desertification, and land use change can lead to displacement, loss of income, and food insecurity.
2. **Frameworks and Tools for Assessment**
    - **Human Development Index (HDI) with Environmental Indicators:** Expanding HDI to include ecological metrics such as ecological footprint and carbon emissions.
    - **Environmental Performance Index (EPI):** Assesses countries on environmental health and ecosystem vitality.
    - **Sustainable Development Goals (SDGs):** Particularly SDG 13 (Climate Action), SDG 14 (Life Below Water), and SDG 15 (Life on Land) intersect closely with human development.
    - **Social-Ecological Systems (SES) Framework:** Explores the interdependence between human societies and ecological systems.

3. **Research Findings and Case Studies**
   - **India's Ecological Development Programs:**
   - *MGNREGA* has contributed to rural ecological restoration while generating employment.
   - *National Biodiversity Action Plan* integrates conservation with development.
   - **Ethiopia's Green Legacy Initiative:** Afforestation campaign aimed at enhancing ecosystem services, improving agriculture, and combating climate change.
   - **Latin America (e.g., Brazil's Amazon region):** Shows how ecological degradationimpacts indigenous populations' health, nutrition, and education access.
4. **Impact of Ecological Factors on Human Development**
   - **Health:** Environmental degradation contributes to respiratory illnesses, vector-borne diseases, and water-borne infections.
   - **Livelihoods:** Ecosystem decline affects agriculture, fisheries, and forest-based economies.
   - **Migration and Conflict:** Environmental stress can lead to displacement and conflict over natural resources.
   - **Education:** Children in degraded environments may have lower school attendance due to poor health or migration.
5. **Policy Implications and Recommendations**
   - **Integrated Planning:** Human development policies must incorporate environmental sustainability.
   - **Community-Based Resource Management:** Empowering local communities to manage natural resources sustainably.
   - **Climate-Resilient Infrastructure:** Developing infrastructure that can withstand ecological shocks.
   - **Ecological Education:** Raising awareness and educating communities on sustainable practices.
6. **Conclusion:** Ecological factors are not just background elements but foundational components of human development. A sustainable future requires aligning environmental health with human development strategies. Interdisciplinary research and policy integration are crucial to ensuring long-term well-being for current and future generations.

Title: On Power and Its Corrupting Effects: The Effects of Power on Human Behavior and the Limits of Accountability Systems
Source: Tobore et al., 2023, Communicative & Integrative Biology (USA)

Introduction

Power is defined as the ability to influence others, dominate resources, and impose one's will asymmetrically. It plays a critical role in human relationships across social, political, and economic spheres. While it can enhance group functioning, it is also known to corrupt—a notion explored historically by thinkers such as Plato and Lord Acton.

Methodology

The article is a **review paper**, synthesizing existing interdisciplinary research from fields such as neuroscience, sociology, physiology, and psychology to examine the effects of power.

Results: The Corrupting Effects of Power

1. Addiction to Power
    - Power can become addictive, driving behaviors like nepotism, elite struggles, and dynastic succession.
2. Moral Hypocrisy and Exceptionalism
    - Power fosters self-righteousness and enables people to excuse their own unethical actions.
3. Disinhibited Behavior and Overconfidence
    - Power leads to increased risk-taking and illusions of control.

      Example: A successful CEO ignores warnings and makes risky decisions due to overconfidence.
4. Unethical Behavior and Entitlement
    - Power encourages self-interest and unethical actions.

      Example: A government official misuses public funds out of a sense of entitlement.
5. Aggressive and Dehumanizing Behavior
    - Power causes individuals to exploit and dehumanize the powerless.

      Example: A prison guard abuses authority, losing empathy for inmates.
6. Sexualization of Social Interactions
    - Increased risk of harassment and abuse due to disinhibited power behavior.

Example: A powerful executive makes inappropriate advances, misusing their position.

7. Hindered Empathy and Cooperation
   - Power reduces perspective-taking and collaboration.

     Example: A manager ignores team input, harming morale and project success.
8. Judgment Bias
   - Power leads to selective information processing and impaired judgment.

Additional Observations

Power Confers Credibility

- The powerful are more readily believed.

  Example: A CEO's risky idea is accepted due to their status.

Power and Victimhood

- Victimhood can be strategically sed to gain sympathy and deflect criticism.

  Example: A leader gains support by portraying themselves as unfairly targeted.

Power and Gossip

- Gossip can be a tool of influence, often weaponized for control or status.

Power and Ambition

- While essential, ambition linked to power can become destructive and unethical.

Power and Accountability Systems

Power imbalances limit the ability of low-power individuals to hold the powerful accountable. Structural issues (e.g., inaccessible or biased courts) often leave abuses unchecked.

Discussion

Power pervades all areas of human life, influencing behavior and ethics. The paper highlights how accountability systems often fail due to structural limitations. It calls for better understanding of psychological and neurological factors to mitigate power abuse.

Conclusion

Power is a double-edged sword—beneficial yet corrupting. Its effects on behavior and the failure of accountability systems demand further research and stronger mechanisms for oversight.

2.class, Gender and Race Intersections in Indian Society in *A Fine Balance* by Rohinton Mistry

Author: Sadia Ashraf, Assistant Professor, Government College Women University, Madina Town, Faisalabad, Pakistan

Source:*Journal of Culture, Society and Development*, 2015

Introduction

This article explores how **class and gender intersect** in shaping individual experiences and social relations. It critiques earlier feminist and Marxist traditions for treating these categories separately, instead advocating for a **relational and intersectional analysis**—highlighting that class is deeply experienced through **gendered and racialized practices**.

Intersectionality Feminist Theory

Intersectionality is a major paradigm in women's studies and textual analysis. It examines how **social identities** (race, gender, class, sexuality, etc.) overlap, creating **unique experiences of privilege and oppression**, enhancing exclusion and marginalization.

Dina Dalal's Story – *A Fine Balance*

Dina Dalal, a key character, symbolizes **intersectional oppression**—through gender, class, caste, age, and widowhood in postcolonial India.

- Orphaned early and dominated by her brother Nuswan.
- Denied education and forced into domestic labor.
- Marries for love; becomes a widow—heavily stigmatized in Indian society.
- Tries to assert independence through tailoring but faces:
  - **Capitalist exploitation**
  - **Landlord discrimination**
  - **Emergency law oppression**
- Her resilience is notable but her life reveals how structures oppress marginalized women.

Review of Literature

The article references theorists like:

- **Kimberlé Crenshaw**
- **bell hooks**
- **Wendy Brown**

It draws on **Marxist and feminist theory**, challenging the illusion of equality in India and examining how **ideologies, institutions, and culture** perpetuate inequality.

Methodology

This is a **conceptual and theoretical** study. The author synthesizes feminist and Marxist literature, analyzing:

- Labor and gender roles
- Domestic labor theories
- Feminist critiques of capitalism

It's a **critical and interpretive approach** examining how theory reflects the lived experiences of working-class women.

Results

Key insights include:

- Gender roles are embedded in class structures.
- Domestic and emotional labor supports capitalism but is unpaid and undervalued.
- Working-class women face double oppression—as workers and as women.
- **Reproductive and care work**, essential to capitalism, is ignored in mainstream economics.
- **Class is gendered**—constructed and experienced differently based on identity.

Intersectional Marxist-Feminist Framework

The article explores how **class, gender, caste, race, age, and nationality** intensify oppression, using **Dina Dalal** as a case study.

Class

From Marxist and Weberian perspectives:

- Class = material and economic conditions
- Bourdieu adds **cultural and symbolic capital**

- Callinicos notes resistance to dismantling class due to **selfishness and greed**

Gender Relations

Patriarchal systems embed power in institutions:

- Men hold power, women are marginalized
- Economic and political exclusion perpetuates **dependence & inequality**

Race

- Adds to gender and class oppression.
- Draws from **Frantz Fanon**—racial and economic hierarchies are intertwined.
- **Dina Dalal's** multiple marginalized identities show how oppression layers within postcolonial, capitalist, patriarchal systems.

Exploitation by the Family Institution

- Dina's brother controls her early life and future.
- Marriage used as a tool for status; widowhood becomes a point of societal shame.

Capitalism and Patriarchy

- Dina seeks independence through work but is exploited (e.g., low wages by Mrs. Gupta).
- **Emergency laws** destroy her fragile autonomy through:
  - Forced evictions
  - State control
- Her oppression is a result of **state, capitalist, and patriarchal forces**.

Public Sphere Exploitation

- Dina faces:
  - Harassment from rent collectors and landlords
  - Legal injustice and sexual harassment
- Her independence is systematically eroded.
- Her life reflects **intersectional oppression** in patriarchal and capitalist society.

Conclusion

The article concludes that:

- A **dual framework (class + gender)** is inadequate.

- A **relational, intersectional approach** is needed to understand how power is co-constructed.
- This framework supports **activism** aimed at dismantling systemic inequality.

3. Human Geographies in Action: Insights into Migration, Development, Culture, and Sustainability by Giuseppe T. Cirella

1. Introduction

Human geography today extends beyond traditional mapping, incorporating **environmental, social, and economic dimensions** to address global challenges. A prominent aspect is **urban design** aimed at achieving ecological balance and human development.

- **Urban green infrastructure** improves public health and offers economic benefits through ecosystem services (e.g., air pollution reduction).
- Cities, particularly **coastal urban areas**, face climate hazards, necessitating **resilience strategies** to tackle these vulnerabilities.

2. Synopsis of Contributions

The Special Issue (SI), titled:

"Human Geography: Interrelationships between People, Place, and the Environment", advances the field through interdisciplinary insights.

Highlights

- Encourages new thinking about spatial and temporal human experiences.
- Structured into **five key themes**:
  - Migration and Population Studies
  - Urban and Rural Development
  - Cultural and Social Geography
  - Political and Economic Geography
  - Environmental and Sustainability Studies

Global research distribution showcases collaborative efforts and geographic diversity, reflecting the increasing importance of **cross-cultural and methodological diversity** in understanding modern human geography.

2.1. Migration and Population Studies

Key Topics

- Displacement

- Return migration
- Intra-urban mobility

Case Studies

- **Nigeria:** Economic incentives foster return of internally displaced persons.
- **China:** Elderly relocation affects mental and physical health.
- **Serbia:** Emotional connection & life satisfaction influence return migration.
- **UK:** GIS tools uncover migration patterns driven by housing price dynamics.

2.2. Urban and Rural Development

- **Urban resilience** relies on **inclusive planning** and **social cohesion**.
- **Rural-urban synergy:** Development in small towns improves rural life.
- **Urban shrinkage** in China and **coastal land loss** in Louisiana highlight urgent need for **adaptive strategies** in planning.

2.3. Cultural and Social Geography

- **Territorial branding** mirrors power structures and regional identity.
- **Heritage regeneration** (e.g., *Tulou* in China, *Marae* in New Zealand) boosts inclusivity and cultural identity.
- **Cultural stereotypes** shape behaviors, such as driving patterns.

2.4. Political and Economic Geography

- **Stock exchange maturity in Africa**linked to sustainable financial growth.
- **Labor market flexibility** underpins economic resilience.
- **Environmental governance:** The EU's biodiversity frameworks guide sustainable policy models.

2.5. Environmental and Sustainability Studies

- Interconnected issues: population, agriculture, urbanization, and health.
- **China:** Ecological strain tied to family planning and land use.
- **Israel & Jordan:** Agricultural risks heightened by climate and elevation.
- **Sustainability requires interdisciplinary strategies** integrating social, economic, and environmental elements.

Introduction

Human development, traditionally assessed through indicators such as health, education, and income, is increasingly being analyzed in conjunction with ecological factors. Ecological factors refer to the natural environment, resource availability, ecosystem services, and the impact of environmental sustainability on quality of life. These components play a critical role in shaping development outcomes, particularly in vulnerable and resource-dependent communities.

2. Key Ecological Factors Influencing Human Development
    - **Natural Resource Availability:** Access to clean water, fertile soil, forests, and biodiversity supports livelihoods, food security, and health.
    - **Climate and Weather Patterns:** Extreme weather events, temperature shifts, and climate change influence agriculture, migration, and public health.
    - **Environmental Quality:** Pollution levels, waste management, and air and water quality directly impact health outcomes.
    - **Ecosystem Services:** Benefits derived from ecosystems, such as pollination, water purification, and climate regulation, support sustainable living.
    - **Land Use and Degradation:** Deforestation, desertification, and land use change can lead to displacement, loss of income, and food insecurity.
3. Frameworks and Tools for Assessment
    - **Human Development Index (HDI) with Environmental Indicators:** Expanding HDI to include ecological metrics such as ecological footprint and carbon emissions.
    - **Environmental Performance Index (EPI):** Assesses countries on environmental health and ecosystem vitality.
    - **Sustainable Development Goals (SDGs):** Particularly SDG 13 (Climate Action), SDG 14 (Life Below Water), and SDG 15 (Life on Land) intersect closely with human development.
    - **Social-Ecological Systems (SES) Framework:** Explores the interdependence between human societies and ecological systems.
4. Research Findings and Case Studies
    - **India's Ecological Development Programs:**
    - *MGNREGA* has contributed to rural ecological restoration while generating employment.

- *National Biodiversity Action Plan* integrates conservation with development.
- **Ethiopia's Green Legacy Initiative:** Afforestation campaign aimed at enhancing ecosystem services, improving agriculture, and combating climate change.
- **Latin America (e.g., Brazil's Amazon region):** Shows how ecological degradationimpacts indigenous populations' health, nutrition, and education access.

5. Impact of Ecological Factors on Human Development
   - **Health:** Environmental degradation contributes to respiratory illnesses, vector-borne diseases, and water-borne infections.
   - **Livelihoods:** Ecosystem decline affects agriculture, fisheries, and forest-based economies.
   - **Migration and Conflict:** Environmental stress can lead to displacement and conflict over natural resources.
   - **Education:** Children in degraded environments may have lower school attendance due to poor health or migration.
6. Policy Implications and Recommendations
   - **Integrated Planning:** Human development policies must incorporate environmental sustainability.
   - **Community-Based Resource Management:** Empowering local communities to manage natural resources sustainably.
   - **Climate-Resilient Infrastructure:** Developing infrastructure that can withstand ecological shocks.
   - **Ecological Education:** Raising awareness and educating communities on sustainable practices.

7. Conclusion

Ecological factors are not just background elements but foundational components of human development. A sustainable future requires aligning environmental health with human development strategies. Interdisciplinary research and policy integration are crucial to ensuring long-term well-being for current and future generations.

Index

P

R

S

U

W

Biodata of Authors

1. Mouneshwari R Kammar

Name and full correspondence address:	:	Dr. Mouneshwari Ramanand Kammar, Professor, Department of Human Development and Family Studies, College of Community Science, Yattinagudda Campus, University of Agricultural Sciences, Dharwad-585005
Email(s) and contact number(s)	:	badigerms@uasd.in, mrkammar13@gmail.com Mob: 9448631170
Institution	:	University of Agricultural Sciences, Dharwad-585005

Academic Qualification (Undergraduate Onwards)

Sl No	Degree	Year	Subject	University/Institution	% of marks
1	B.H.Sc.	1994	Home Science	College of Rural Home Science, UAS, Dharwad	89.60%
2	M.H.Sc.	1997	Human Development	College of Rural Home Science, UAS, Dharwad	91.20%
3	Ph.D.	2004-2009	Human Development	Smt Nathibai Damodar Thackersey Women's University, Mumbai	Satisfactory
4	MBA	2009	Human Resource Development	Institute of Business Management, Hubli	Satisfactory
5	PGDND	2012	Nutrition and Dietetics	Karnataka State Open University, Mysore	72.50%
6	PGDAEM	2013	Agri. Extension Management	MANAGE, Hyderabad	79.36%

Ph.D thesis title, Guide's Name, Institute/Organization/University, Year of Award

Title: Changes in Ageing: A cohort Study

Guide's Name: Dr. Ravikala Kamath

Institute / University: Smt Nathibai Damodar Thackersey Women's University, Mumbai

Year of Award: 2009

Work experience (in chronological order).

Sl No	Positions held	Name of the Institute	From	To	Pay Scale
1	Field Enumerator	UAS Dharwad	February 1997	May1997	Contractual basis
2	Research Associate	College of Rural Home Science, UAS Dharwad	June 1997	13.5.2005	8800+HRA
3	SMS-Home Science	BIRDS Krishi Vigyan Kendra, Tukkanatti	30.5.2005	16.2.2009	8000-13500 pay scale
4	SMS-Home Science	Krishi Vigyan Kendra Raichur, UAS Raichur	21.2.2009	16.2.2017	15600-39100 with AGP of. Rs.6000
	SMS-Home Science	Krishi Vigyan Kendra Bagalkote, UAS Dharwad	17.2.2017	24.7.2017	15600-39100 with AGP of. Rs.6000
5	Senior scientist and Head in the scale of Associate Professor	Krishi Vigyan KendraBagalkote, UAS Dharwad	24.7.2017	4.3.2023	131400 as basic pay
6	Associate Professor	College of Community Science, UAS Dharwad	5.3.2023	Till date	152300 as basic pay

Professional Recognition / Award/ Prize / Certificate, Fellowship received by the applicant

Sl No	Name of Award	Awarding Agency	Year
1	Choudary Devilal award for outstanding AICRP award received as Team member in AICRP which was adjudged as the Best AICRP Team award	Indian Council of Agricultural Research, New Delhi	2003
2	Scientist of the year award	National Environmental Sciences, Academy New Delhi	2011
3	Best oral presentation for Drudgery Reducing Equipment as key to enhance the efficiency of farm women presented at National Seminar on Attracting farm youth to sustainable agriculture held at GKVK on Aug. 26-28, 2011	University of Agricultural Sciences, Bengaluru	2011
4	Overall performance of KVK during for as in charge of KVK Raichur	Agriculture Research Application and Research Institute, Bengaluru	2015

5	International conference on Extension for strengthening Agricultural Research and Development held at JSS Sutturu, Mysuru. December 14-16, P:108	ICAR-JSS KVK Suttur, ICAR-ATARI, Bengaluru; Extension Education Society (EES) and Tamil Nadu Agricultural University (TNAU)	2019
6	Pandit Deen Dayal Upadhyay Krishi Vigyan Rashtriya Protshahan Puraskar 2019- Best KVK award for zone -11	Indian Council of Agricultural Research, New Delhi	2019

Publications (List of papers published in SCI Journals, in year wise descending order

Sl. No.	Authors	Title of the paper	Journal and details	Page	Year
1	**Kammar M.R.** and Vanishree S.	Weeding tools- a source of drudgery reduction for farm women.	Multilogic in Science. 9(31):156-157 NAAS Rating 5.20	156-157	2019
2	**M R Kammar,** R Kamath, KV Ashalatha	Correlates of functional abilities among aged	Karnataka journal of agricultural sciences 23 (3) : (480-484) 2010 NAAS Rating 4.42	480-484	2010
3	**Kammar , M R,** Kamath and Ashalath K V	Functional abilities of the aged	International Journal of Gerontology 25(1): 41-50	41-50	2011
4	**Badiger M S,** R Kamat and Ashalatha K V	Socio-demographic attributes of the elderly : a gender perspectivel	Advance Research Journal of Social sciences, 2(1): 1-4	1-4	2011
5	Vijaykumar Ilkal, Jagruthi D , Hugar L B, S S Patil and **Kammar M R**	Constraints faced by farm women in dairy farming	Research Journal of Agricultural sciences, 4 (5/6): pp688-690 NAAS Rating 4.54	688-690	2013
6	Ravula Sudharshan Reddy, CT Ramachandra, Sharanagouda Hiregoudar, Udaykumar Nidoni, Jagjiwan Ram, **Mouneshwari Kammar**	Influence of processing conditions on functional and reconstitution properties of milk powder made from Osmanabadi goat milk by spray drying	Small ruminants research 119(2014): 130-137 NAAS Rating 7.21	130-137	2014

Sl. No.	Authors	Title of the paper	Journal and details	Page	Year
7	Rashida Rahman, Sharanagouda H. M. Veeranagouda, c. T. Ramachandra, **Mouneswari Kammar,** Udaykumar nidoni and R. S. Roopa	Physico-chemical, textural and sensory properties of muffins fortified with wheat grass powder	*Kar J Ag Sc*, 27(4): 553-554 NAAS Rating 4.42	553-554	2015
8	**Mouneshwari R** Kammar and Pramod Katti .	On Farm Testing on Assessment of Different Type of Weeders in Direct Seeded Rice	*J. of Krishi Vigyan.*, 3(Special Issue) : 70-74 NAAS Rating 4.41	70-74	2015
9	Shakuntala N.M. **Mouneshwari Kammar,** Vasudevan S N, Biradar S A and Rajeshwari S Alur	Value added products from Linseed	Progressive Research 11(Special Issue): 5634-5638 NAAS Rating 3.84	5634-5638	2016
10	Thasnimol, H Lokesh, **M.R. Kammar** , Amrutha Joshi and Siddayya	Consumption of High value Agricultural commodities in North Eastern Karnataka-An economical analysis	Progressive Research- An international Journal 11(Special issue-Part III): 719-727 NAAS Rating 3.16	719-727	2016
11	Vanishree S and **Kammar M R**	Factors influencing dietary habits of preschoolers	Advances in life sciences 5(17):6609-6612 NAAS Rating 3.15	6609-6612	2016
12	Vanishree S , **Kammar M R** and Bhavani	Suitability of handgloves for harvesting chickpea	Journal of farm sciences, 29(5): 828-89 NAAS Rating 4.42	828-89	2016
13	Vanishree S , Kammar M R and Nidoni U	Development and evaluation of pearlmillet based novel health drink	Advances in life sciences, 5(13):5483-5486 NAAS Rating 3.15	5483-5486	2016

Sl. No.	Authors	Title of the paper	Journal and details	Page	Year
14	Thasnimol M. Lokesh **Kammar , M R** Amritha Joshi and Siddayya	Consumption pattern in Karnataka with reference to high value agricultural commodities	International Journal of Applied and pure sciences in Agriculture P 132-138, 2(3):133-138	132-138	2016
15	Thasnimol M. Lokesh **Kammar , M R**	Consumption pattern in India with reference to high value agricultural commodities	Advances in Life sciences, 5(6): 2242-2246 NAAS Rating 3.15	2242-2246	2016
16	K.Y. Ullikashi1, M.R. Kammar and Sudeep R. Lokapure	Development of Value Added Products from Bael Fruit (Aegle marmelos)	International Journal of Current Microbiology and Applied Sciences. 6 (7): 2652-2659 NAAS Rating 5.28	2652-2659	2017
17	**Mouneshwari R Kammar,** Vidyavathi G Y and Amaresh Y S.	. Effect of Demonstration on Use of Paddy Straw Baler in Raichur District.	J Krishi Vigyan 2017, 6(1) 224-226. NAAS Rating 4.41	224-226	2017
18	Majeed G, **Mouneshwari Kammar** and Angadi S.C.	Analysis of Kisan Mobile Advisory Text message sent to farmers of Bagalkot district	Int. Journal of current Microbiology and applied sciences, (2018):7(7) Pp: 1241-1246 NAAS Rating 5.28	1241-1246	2018
19	**Kammar M.R.**, A.P. Biradar, Arjun Sulagitti, S. Sudha.	Self sustainability of farming community through promotion of organic agriculture:A case study of KVK, Bagalkot, India.	*International journal of current microbiology and applied sciences*, 7(9):3653-3658 NAAS Rating 5.38		2018
20	**Kammar M.R.** and S. Vanishree	Supplementation of Low Cost Alternative Foods to Combat Vitamin C Deficiencies among Children.	*International Journal of Current Microbiology and Applied Sciences,* 7(8): 4143-4146, NAAS Rating 5.38		2018

Sl. No.	Authors	Title of the paper	Journal and details	Page	Year
21	**Kammar M.R.** , A.P. Biradar, Arjun Sulagitti, and Mahesh Khadagi.	An Experience of Hydroponics Fodder Production by Farmers of Bagalkot district	*Journal of Pharmocology and Phytochemistry January 2019, 8(1): 1033-1035,* NAAS Rating 5.28	*1033-1035*	2019
22	**Kammar M.R.** Dinesh Kumar SP and Angadi S C.	Potential future technology: Mechanized harvesting of sugarcane.	Journal of Pharmacognosy and Phytochemistry. 8(4): 1520-1522 NAAS Rating 5.28	1520-1522	2019
23	**Kammar M.R.,** Kadagi M., Sulagitti A.R. and Majeed G.	Assessment of methods for nutritional adequacy in agrobased farming system.	International Journal of Recent Scientific Research. 10(C): pp. 35312-35314 NAAS Rating 3.65	35312-35314	2019
24	**Kammar M R** , Arjun, Mahesh K and A P Biradar	Experience of Hydroponic fodder production by farmers of Bagalkote district	Journal of Pharmacognosy and Phytochemistry. 1(8):1033-1035 NAAS Rating 5.28	1033-1035	2019
25	**Kammar M R** , Arjun S and Angadi S C	Entrepreneurship development of millets and jaggery	Journal of Pharmacognosy and Phytochemistry. 9(2):1030-1032) NAAS Rating 5.28	1030-1032	2020
26	Mahesh Kadagi, **Kammar M R,** Ashoka P	Impact of sheep and goat rearing on skill training on knowledge gain and adoption of technologies	Int. Journal of current Microbiology and applied sciences, 2020, 9(4) Pp:2144-2151 NAAS Rating 5.28	2144-2151	2020
27	**Kammar, M R ,** Angadi S C and Dineshkumar	Demonstration of nutria farms for year round nutrition security among farm families	Int. Journal of current Microbiology and applied sciences, 2020, 9(6) Pp:2350-2355 NAAS Rating 5.28	2350-2355	2020
28	Dineshkumar S P, **Kammar M R** and Sudah S	Performance of dicoccum wheat DDK 1029 under FLD in Bagalkot district of Karnataka	International Journal of Chemical studies , 9(1): 1390-1392 NAAS Rating 5.31	1390-1392	2021

Sl. No.	Authors	Title of the paper	Journal and details	Page	Year
29	Sudha S, Kammar M R, Dineshkumar S P and Airadevi Angadi	Integrated Disease Management for the control of bacterial blight in pomegranate in Karnataka	Journal of Krishi Vigyan December 10(1): pp: 118-122 NAAS Rating 4.55	118-122	2021
30	Siddappa C Angadi, Chidanand P Mansur and Mouneshwari R Kammar	Review of indigenous technology knowledge on soil and water conservation	The Pharma Innovations. The Pharma Innovation Journal 2021; SP-10(12): 1285-1289 NAAS Rating 5.23	1285-1289	2021
31	Siddappa C Angadi, Chidanand P Mansur and Mouneshwari R Kammar.	Indigenous technical knowledge on soil and water conservation, soil health management and tillage techniques in Bagalkot district..	The Pharma Innovation Journal 2021; SP-10(12): 1290-1292. NAAS Rating 5.23	1290-1292	2021
32	Dinesh Kumar S P, Sudha S and Kammar M R	Performance of Cowpea Variety DC-15 at Farmers' Field.	J Krishi Vigyan 2022,10 (2) : 263-267. NAAS Rating 4.47	263-267	2022
33	Mouneshwari R Kammar, Arjun R Sulagitti and Sudha S	Performance of biopesticides for management of white grub *Holotrichia serrata* in sugarcane.	The Pharma Innovation Journal 2022; 11(11): 2152-2154. NAAS Rating 5.23	2152-2154	2022
34	Mouneshwari Kammar, Angadi SC, Dineshkumar SP and Arjun RS	Impact of cluster front line demonstrations on yield levels in Pigeonpea and chickpea.	2023The Pharma Innovation Journal .12(3): 306-309 NAAS Rating 5.23	306-309	2023

Books/Reports/Chapters/General articles etc.

Success stories of farmers Published by researcher in popular magazines

Sl No.	Authors	Magazine, Vol and Page No.	Title in English
1	Kurubar A. R. and **Kammar M.R.**	Sharad krishi, March 2011: 22-24	Turmeric cultivation in Paddy conventional area
2	Kurubar A. R., **Kammar M.R.** and Chanvan M.	Annadata Marh 2011: 9-11	Success story of an IFS Farm woman:Sheela
3	Kurubar A R and **Kammar M.R.**	Sharad krishi, April 2011: 34-35	Mustard : an alternate crop to summer paddy in Raichur district
4	Kurubar A. R., **Kammar M.R.** and Chanvan M.	Sharad krishi, July 2011:26-28	Success story of a woman Sweet orange grower
5	Kurubar A. R., **Kammar M.R.** and Chanvan M.	Sharad krishi, August 2011: 7-926-28	Success story of a jasmine cultivator
6	**Kammar M.R.,** Pramod Katti and Chavan	Sharad krishi, Oct 2012:52	Success story of rural youth entreneur
7	**Kammar M.R.,** Udaykumar Nidoni and Geeta H P	Krishi Pradeepike July-September-2014:32-35	Success story of Millet farmer: Kallappa
8	**Kammar M.R.,** Ramesh G and Pramod Katti	Krishi Pradeepike Jan-March 2015:48-50	Success story of a Bhendi farmer
9	**Kammar M.R.**, Ramesh G	Krishi Pradeepike Oct-Dec 2015:46	Success story of an IFS youth farmer
10	**Kammar M R** and A P Biradar	Krishi Munnade: 30(6) pp34-35 2017	Success story of a farmer with animal husbandry
11	**Kammar M R** and A P Biradar	Krishi Munnade: 30(7) pp31-32 2017	Success story of pomegranate Farmers association
12	**Kammar M R**	Krishi Munnade: 31(4) pp 28-29 2017	Success story of organic farmer Sri Dhareppa Kittur
13	**Kammar M R** and Dineshkumar S P	Krishi Munnade: 31(8) pp 34-342018	Success story of a female youth graduate
14	**Kammar M R**	Krishi Munnade: 31(10) pp 34-35 2018	Success of farmer from joint family
15	**Kammar M R** and Mahesh Kadagi	Krishi Munnade: 31(10) pp 34-35	Innovative farmer of Shri Dundappa Halli
16	**Kammar M.R.**	Krishi Munnade: March 2020, 33(3) : pp 32	Success story of a woman farmer
17	**KVK Bagalkot**/FPO Hunagund	Innovative solutions during COVID-19 Page 50-51	Success story of contact FPO of KVK bagalkot which was technically supported durig COVID 19

Sl No.	Authors	Magazine, Vol and Page No.	Title in English
18	**KVK Bagalkot**/FPO Hunagund	Innovative solutions during COVID-19 Page 50-51	Success story of contact FPO of KVK bagalkot which was technically supported durig COVID 19
19	**Kammar M R and vijayamahantesh**	Vijaya Karnataka Daily paper 14.09.2020	Innovative farmer who cultivated paddy in sugarcane conventional area
20	**Kurubar A R and Kammar M R**	Sharad Krishi Sept .2011: 37-39	A success story of farmers planning crops according market demand
21	**KVK Bagalkot**	Innovations of farmers-2020 published by UAS Dharwad P: 3	Success story of IFS farmer Shri Basayya Ningayya Salimath
22	**KVK Bagalkot**	Innovations of farmers-2020 published by UAS Dharwad P: 4	Success story ofchilli farmer (Small farmer): Shri Hanamant Siddappa Garagad
23	**KVK Bagalkot**	Innovations of farmers-2020 published by UAS Dharwad P: 4	Automatic waterer to animals: Shri Muttanna Bairamatti
24	**KVK Bagalkot**	Innovations of farmers-2020 published by UAS Dharwad P: 5	IFS farmer from Amaravathi: Shri N M Sandigwad
25	**KVK Bagalkot**	Innovations of farmers-2020 published by UAS Dharwad P: 5	Innovation of youth farmers Shri Ramesh Hadagali
26	**Mouneshwari R. Kammar** and Mahesh Kadagi	RiImagining Agriculture-Farmers Innovations Editors: Chandregouda M.J., DVSrinivasa Reddy, Randhir Singh, Mallikarjuna B. Hanji, M.R.Kammar, N Loganandan, Bharati Hegde 2019 Published by Director, ICAR ATARI, Bengaluru 2020	Intercropping hellos in Eco-friendly Pest Management.
27	Kammar M R and Arjun R S.	1. Vijaya Karnataka 28 , June 2021	Success story of farmer exporting banana to Arab countries
1	Kammar M R and Arjun R S.	Vijaya Karnataka 21, June 2021	Success story of farmer with scarcity of water in agricutlure

Sl No.	Authors	Magazine, Vol and Page No.	Title in English
2	Arjun R S and Kammar M R	Krishi Kamadhenu April 2021 P: 21-23	Success story of a physically diabled person in agriculture
3	M R Kammar	Krishi Kamadhenu,December 2021. Dec 2021. 35	Success story of joint family which has succeeded in agriculture
4	Arjun R S and Kammar M R.	Vijaya Karnataka 17 May 2021	Success stories of farmers cultivating multiple crops
5	Mouneshwari R Kammar.	Vijaya Karnataka 25 october 2021	Success story of women with very small landholding practicing organic agriculture
6	Kammar M R	Krishi Kamadhenu Monthly February 2022	Mangement of snails and slugs
7	Mouneshwari Kammar	Vijaya Karnataka April 11, 2022 Page 8	Success story of farmers Basavaraj Bennur adopting agri technology for enhancing income
8	Nagaleekar B., Kammar M R and Lamani Sunil	Krishi Kamadhenu June 2022 15(6):23-25	Introduction of Mousam app
9	Arjun R S and Kammar M R	Krishi Kamadhenu June 2022 15(6):30-33	Cultivation practices of soybean
10	M R Kammar and Nagaleekar	Vijayakarnataka 11.7.2022 Page – Krishi Vijaya	Success story of manjunath Dalin who has adopted cultivation of varieity of guava
11	Nagaleekar B, Kammar M R and Sunil Lamani	Krishi Kamadhenu15(7): 19-22	Need for agriculture based on weather forecasting
12	M R Kammar and Nagaleekar	Vijayakarnataka 25.7.2022 Page – Krishi Vijaya	Success story of Dnyaneshwar Mundinamani
13	S C Angadi and M R Kammar	Vijayakarnataka 22.11.2022	Success story of farmer adopting green manuring in sugarcane
14	Nagaleekar B, Kammar M R and Sunil Lamani	Krishi Kamadhenu December 2022 28-30	Management of Ozone layer
15	Mouneshwari R Kammar	Krishi Kamadhenu 16(1): January 2023 35-36	Introduction to nutrigarden
16	Nagaleekar B, Kammar M R and Sunil Lamani	Krishi Kamadhenu16(1): January 2023 37-38	Introduction to Meghdoot app

Booklets/Training manuals prepared

Sl No.	Title	Authors	Prepared for
1	A strategy document on technological interventions for improving agricultural scenario in Bagalkot district	M R Kammar, Dineshkumar S P., Arjun Sulagitti, Mahesh Kadagi, Sudha S., Airadevi Angadi, S C Angadi and Majeed G	A KVK Bagalkot Document prepared during 2019-20
2	Quinqunennial review team report 2011-12 to 2018-19	M R Kammar and team	A KVK Bagalkot Document prepared during 2020
3	Training manual on production of organic manures	Arjun Sulagitti, M R Kammar, Dineshkumar S P and other	Training manual for trainees; KVK Bagalkot Document
4	Training manual on production of organic nputsmanures	Arjun Sulagitti, Venkanna Balganur, M R Kammar, Siddappa Angadi , Airadevi Angadi, Dineshkumar S P and other	Training manual for trainees; KVK Bagalkot Document
5	Training manual on varieties of nutrigardens	Airadevi Angadi, Kammar M R , Kumar H, Dineshkumar S P and Others	Training manual for trainees; under nutrigardens:KVK Bagalkot Document
6	Dairying and Vermicomposting	A P Biradar, Dineshkumar, Yadahalli, Hugar S V and M R Kammar	Training manual on organic agriculture and vermicomposting
9	Coffee table book of KVK Bagalkote	Kammar M R and others	Showcasing the achievements of KVK Bagalkote
10	Compilation of 110 case studies of farmers of Bagalkote district whose income is doubled during 2018-19 to 2021-22	Kammar M R and others	DFI document of KVK Bagalkote submitted to ATARI Bengaluru
11	Meghdoot newsletter	DVS Reddy , Kammar M R and others	Newsletter of DAMU units of Zone XI , ATARRI Bengaluru
12	Handbook published as part of NABARD sponsored training sanctioned to KVK Bagalkote	Airadevi Angadi, Shripad Kulkarni, Mouneshwari Kammar. 2022	. Handbook on Processing and Value Addition to Fruits and Vegetables.UASD Publication with ISBN no. 978-81-88367-69-9,

List of folders relevant to Extension work

Sl No	Authors	Year of publication	Title of Leaflet/Folder
1	**Kammar M R** and Kurubar A R	2010-11	Millets for Health
2	Vanishree S, **Kammar M R**	2011-12	Iron Deficiency Anemia-An information
3	**Kammar M R** , Vanishree S and Shilpa D H	2012-13	Balanced diet for children and women
4	Benki A M, **Kammar M R** and Chavan M	2011-12	Helathy millet-Foxtail millet
5	**Kammar M R** , Vanishree S Roopabai S and Benki A M	2013-14	Value addition : as an Entrepreneurial activity
6	Vanishree S **,Kammar M R,**Nidoni U and Upperi S N	2012-13	Value added products from Bajra
7	Kammar M R and Katti P	2013-14	Spiral separator for clean and quality redgram
8	**Kammar M R** , Katti Pramoda, Shakuntala N M , Vanishree S and Sangeeta M	2014-15	Conventional methods of grain storage
9	**Kammar M R** and Vanishree S	2014-15	Smokeless Chula with agriculture residues
10	**Kammar M R** , Anupama C, Benki A M and Mohan Chavan	2014-15	Value addition to Bajra
11	**Kammar M R ,** Chanvan M, Benki A M and Vanishree S	2014-15	Tamarind for Entrepreneurship Development
12	Udaykumar Nodoni,**Kammar M R**, Ambarish Ganachari ad Mathad P F	2014-15	Post Harvest Technology of Turmeric and its value addition
13	Sujay H, AMaresh Y S, Chavan M, Prabhuraj and **Kammar M R**	2015-16	Management of Bee hives during summer
14	Chavan M**, Kammar M R** AMaresh Y S	2015-16	Production technologies in groundnut
15	Vidyavathi Yadahalli, Chavan M and **Kammar M R**	2015-16	Importance of soil testing and soil health card
16	Shakuntala N M, Jyothi vastrad, Vasudevan S N **kammar M R**	2016-17	Handmade paper from LinseedFibre
17	Geeta H P, **Kammar M R**, P F Mathad *et al.*	2014-15	Post Harvest Technology and value addition to millets
18	**Kammar M R**	2015-16	Drudgery reducing equipments for women in dryland agriculture
19	Published by **Kammar M R**	2015-16	Integrated Farming System
20	**Kammar M R ,** Sudha S Dineshkumar S P	2019-20	Nutrigarden for household nutrition security

Sl No	Authors	Year of publication	Title of Leaflet/Folder
21	Published by **Kammar M R**	2019-20	Efective utilization of Farm pond
22	Airadevi *et al* Edited by M R Kammar and published by Belli R B.	2021	Khuski Totagarike
23	Sudha *et al* Edited by M R Kammar and published by Belli R B.	2021	Management of disease s in onion
24	Arjun *et al* Edited by M R Kammar and published by Belli R B.	2021	Management of root grub in sugarcane
25	Arjun *et al* Edited by M R Kammar and published by Belli R B.	2021	Snails and slug management
26	Dineshkumar *et al* Edited by M R Kammar and published by Belli R B.	2021	Management of striga in sugarcane
27	Dineshkumar *et al* Edited by M R Kammar and published by Belli R B.	2021	Contingent crop planning for Bagalkote district
28	Airadevi *et al* Edited by M R Kammar and published by Belli R B.	2021-22	Value addition to Banana
29	Nagaleekar *et al* Edited by M R Kammar and published by Belli R B.	2022-23	Introduction of Mousam app
30	Kammar M R *et al*	2022-23	Value addition of Jaggery
31	Sudha *et al* Edited by M R Kammar and published by Belli R B.	2022-23	Integrated Disease management in Pigenpea.
32	Airadevi *et al* Edited by M R Kammar and published by Belli R B.	2022-23	Importance and uses of Micro nutrients in Horticultural crops.
33	Balaganur V *et al* Edited by M R Kammar and published by Belli R B.	2022-23	Importance of Green fodder in successful dairy.
34	Nagaleekar *et al* Edited by M R Kammar and published by Belli R B.	2022-23	Need for Weather based agriculture
35	Nagaleekar *et al* Edited by M R Kammar and published by Belli R B.	2022-23	Mobile applications for weather forecasting information
36	Airadevi *et al* Edited by M R Kammar and published by Belli R B.	2022-23	Value addition to Banana
37	Nagaleekar *et al* Edited by M R Kammar and published by Belli R B.	2022-23	Introduction to Mausam App

Any other Information (maximum 500 words)

I Dr. Mouneshwari Ramanand Kammar, was working as Senior Scientist and Head, ICAR-Krishi Vigyan Kendra, Bagalkote , University of Agricultural Sciences, Dharwad has worked on issues related to women empowerment enhancing the capacity of women to make choices and to transform these choices into desirable actions and outcomes. The major areas under study were drudgery reduction enabling the women to access the technologies, facilitating them to have access to fuel needs of rural family, access to income and entrepreneurship development opportunities for women and rural folk, access to nutritious food and health by establishing nutrigardens for the year round supplementation of nutrients to farm families and schools, , access to food grains at households by implementing grain storage studies, ensuring nutritional security of farm families. The interventions on mechanization in major commercial crops such as paddy and sugarcane are also notable. The researcher has used various innovative methodologies including social media, TV/Radio, popular articles and other print and electronic media. Functional linkages established for the welfare of women in rural area established a strong base for empowering women. The outcomes of the work submitted for the award have been implemented in the form of policies. Though small, her works have given scope for commercialization of technologies in the area of processing and value addition. Impact analysis of the technologies imparted conducted at KVK Bagalkote imply the need to further expand the intervention to the unreached. Documentation of case studies and research papers emerging out of extension work deserve appreciation. She is involved in disseminating technologies which are relevant to existing cropping system of the district in collaboration with various departments and farmers. The experience gained through extension work in three districts representing north Karnataka has benefitted the area in which she is working. As Associate Professor of Human Development and Family Studies, also I am executing the duties to the best of my abilities.

2. Dr. Vinutha U. Muktamath, Scientist, AICRP-WIA, UAS, Dharwad

Expertise available with the investigators in executing the project

A human development specialist trained in assessment techniques for young children, adults and elderly and counselling especially in cognitive behaviour therapy. Have proficiency in qualitative research methodology, infant and child studies, developmental challenges and theories of human behaviour. Guided four Ph.D. students and six MSc students. Developed technological packages on Postnatal care of mother & children, Successful breastfeeding practices and quality of life of elderly. Published more than 35 research papers, 15 books, and chapters in books and manuals of national and international repute. Attended various national and international conferences and presented research papers and won best research paper award for 6 papers in various national & international seminars. Won best thesis award for the Doctoral work on "Breast feeding practices and its influence on infant development outcomes" namely cognitive, psychomotor, language and socio-emotional development in 1st Indo-Asean Conference on Innovative Approaches in Applied Sciences and Technology held at Nong Lam university, Ho Chi Min city Vietnam.

Conducted research on ADHD prevalence and risk factors among children and Cerebral palsy in North Karnataka region as well as externalising and internalising problems in children which throws the light on behavioural issues in children and the need to use digital media in a right way to tackle these issues.

Key publications published by the Investigators pertaining to the theme of the proposal during the last 5 years

1. **Muktamath, V.U and** U.C. Muktamath, 2016, Comparitive Analysis of Anthopometric Variable among three different races (Mangoloids, Negroids and Local Indian College Men Students Research, Reaction and Resolution, 4(1): 33-40
2. **Muktamath, V.U.,** Khadi, P.B and Muktamath, U.C., 2016, Interrelation of emotional intelligence and general intelligence among staff and students of UAS, Dharwad, International Journal of Humanities and Social Sciences, 5(5): 95-100
3. Veena S. Jadhav, Yogita Masur & **Vinutha U. Muktamath, 2016,** Health Hazards of women and men workers in building construction, Journal of Farm Science. Special Issue, 29(5): 613-616
4. Patil, S., Khadi, P.B **and Muktamath,V.U**., 2016, **Influence of home environment on urban and rural infants physical growth statu,** International Journal of Humanities and Social Sciences, 5(5): 101-106

5. **Muktamath, V.U.** and Itagi, S., 2017, Socio-cultural determinants of breast feeding practices in Dharwad district of Karnataka, Advance *Research Journal of Social Sciences*, 8(1): 23-29
6. **Muktamath V.U.**, Itagi S.K and Khadi, P.B., 2018, Breast feeding knowledge, attitude and practices: A comparative study of urban and rural mothers in northern Karnataka, India. *Asian J of H.Sc*, 13(1): 55-61
7. **Muktamath V.U.** and Itagi S.K., 2019, Breast feeding practices: Effect on infant's survival, health and development, *J of Expt. Zool,* 22(1): 239-244
8. **Muktamath V.U.** and Itagi, S.K., 2019, Breast feeding practices: Effect on infant morbidity. *International Journal of Chemical studies*, 7(6): 1028-1033
9. Koppad, A. and **Muktamath V.U.,** 2019, Functional status among rural and urban elderly in Ranebennur taluk, Haveri District, Karnataka, India. *International Journal of Current microbiology*, 18(9): 8-14
10. Koppad, A. and **Muktamath V.U.,** 2019, Prevalence of geriatric depression among community dwelling elderly, India. *International Journal of Chemical studies*, 7(6): 297-301
11. Koppad, A. and **Muktamath V.U.,** 2019, Predictors of functional status among community dwelling elderly. *International Journal of Chemical studies*, 7(6): 288-293
12. **Muktamath V. U,** Khadi P.B., Hegde P.R and Koppad, A., 2020, Nutritional status of Elderly of Dharwad district: An intervention study. *International Journal of Chemical studies*, 8(6): 815-818
13. Massar D., Patil P.B., Pujar L. and **Muktamath V.U., 2021**. Parenting Practices as Predictors of Adolescents Externalizing and Internalizing Problems, *Journal of Community Mobilization and Sustainable Development* 16(1): 1-6 5.3 NAAS
14. **Dr. Vinutha U. Muktamath**, Dr. Pushpa B Khadi and Priya Ramesh Hegde., 2021, Nutritional status and risk factors associated among rural elderly in Northern Karnataka: An intervention study. The Pharma Innovation Journal., SP-10(12): 1101-1107. NAAS-5.23
15. **Vinutha U. Muktamath**, Sadhana Kulloli, Priya R Hegde (2022). Postnatal Care of Mothers: Knowledge, Nutritional Status and Utilization of Postnatal Care Services. J Edu Psyc Res, 4(1): 312-319 Impact factor-0.8, NAAS- 6.
16. Samreen Chand, **Vinutha U. Mukthamath**, Lata Pujar and Geeta

Channal, 2022, Prevalence postpartum depression among women in Bagalkot district of Karnataka, Journal of Farm Science, 35(2): 282-28 **NAAS: 4.42**

17. **Vinutha U. Muktamath**, Sunanda Itagi and Pushpa B Khadi, 2022, Breast feeding Experience of Women – The essence of Womanhood: A Qualitative Study, Journal of Community Mobilization and Sustainable Development. 3: 946- 95 **NAAS-5.67**
18. Mukta G. Sthavarmath, Lata Pujar and **Vinutha Muktamath**., 2022, Parent child relationship and social – cognition among pre – school children. The Pharma Innovation Journal, 11(10): 2093-2098 **NAAS: 5.23**
19. Priya R. Hegde and **Vinutha U. Muktamath**, 2022, Health status of Halakki Vokkaliga and Kunabi tribal women in Uttara Kannada district of Karnataka. The Pharma Innovation Journal, 11(11): 2649-2652 **NAAS: 5.23**
20. Sunitha N.H., Lata Pujar and **Vinutha U. Muktamath.,** 2022, Influence of parental encouragement on academic achievement of college students, The Pharma Innovation Journal 2022; SP-11(6): 2152-2156
21. **Vinutha U. Muktamath** and Sunanda K Itagi, 2023. Effect of Breastfeeding Practices on Psychomotor Development of Infants. Biomed J Sci & Tech Res 49(5): BJSTR. MS.ID.007856.
22. Shweta Biradar and **Vinutha U. Muktamath**, 2023, Scholastic difficulty: scale Development and Implications for future Research,Asian Journal of Agricultural extension, Economics, and Sociology, 41(6): 17-27, ISSN: 2320-7027
23. Shweta Biradar and **Vinutha U. Muktamath**, 2023, scale development to measure coping mechanisms among pre-university college students, The Pharma Innovation Journal, 12(4): 2078-2086
24. Rekha and **Vinutha Muktamath**, 2023, Influence od socio-demographic factors on cerebral palsy among children. Journal of farm science, 36(1): 109-112
25. Rekha and **Vinutha Muktamath**, 2023, Factors influencing personality traits of rural and urban college students.Journal of farm science, 36(1): 113-118

Books Published

1. Khadi P.B.,Badami,Sand Muktamath,V.U,2016,Innovative Stragies for quality life of elderly,Directorate of Research, UAS, Dharwad

2. Vinuta U Muktamath , Sunanda Itagi and Sadhana Kullolli,2018, Postnatal Care of Mother and Newborn,(English)Directorate of Research & KVK, Dharwad, UAS, Dharwad
3. Vinutha U.Muktamath&Sunanda Itagi, 2018,Effect of prenatal maternal stress on infant development, Lambert Academic publishing, Mauritus, ISBN:978-613-9-87301-2
4. Vinutha U.Muktamath. Sunanda Itagi, Ramita& Sadhana Kulloli, 2019, Postnatal care of mother & new born (kannada), Directorate of Research, UAS, Dharwad ISBN:978-81-88367-46-7
5. P.B.Khadi,Vinutha.U.Muktamath. Sumagala Badami & Priya Hegde,2019, Problems faced by elderly and its management(Kannada), Directorate of Research ,UAS, Dharwad
6. P.B.Khadi,Vinutha.U.Muktamath. Priya Hegde &Sumagala Badami,2019, Nutrional care for elderly(Kannada),, Directorate of Research ,UAS, Dharwad
7. P.B. Khadi, Manjula Patil, Priya Hegde , Sumagala Badami & Vinutha. U.Muktamath, 2019, Mental health issues of elderly(Kannada), Directorate of Research, UAS, Dharwad
8. Saraswati Hunshal,Vinutha.U.Muktamath&Shrila Matti, Safe Motherhood, (Kannada), Directorate of Research ,UAS, Dharwad
9. Vinutha U.Muktamath ,Sunanda Itagi&Aishwarya Koppad, 2019,Breast feeding & cognitive development,Lambert Academic publishing, Mauritus ISBN:978-620-0-48337-9
10. Vinutha.U.Muktamath. Aishwarya Koppad& Priya Hegde,2019, Cogitive issues in elderly(Kannada),Directorate of Research, UAS , Dharwad
11. Rekha Rajkumar and Vinutha Muktamath, 2023, Cerebral Palsy among Children, Lap Lambert Publishing company, U.K, ISBN-978-620-5-63256-7
12. Vinutha U. Muktamath, Sunanda K Itagi & Priya R Hegde., 2021, Human Milk Composition: Potential Role in Infant Health and Disease, INSC International Publishers.
13. Priya Hegde and Vinutha U. Muktamath, 2022, Suicidal Thoughts and Behaviour among Adolescents. LAP LAMBERT Academic Publishing. ISBN: 978-620-4-73971-

Book Chapter

1. **Vinutha U. Muktamath**, Priya R. Hegde and Samreen Chand L., Types of Specific Learning Disability., Learning Disablities, Intech Open Publishers London. ISBN 978-1-83968-588-0.
2. Rekha Rajkumar **and Vinutha Muktamath,** 2023, Cerebral Palsy among Children, Lap Lambert Publishing company, U.K, **ISBN-978-620-5-63256-7**
3. Koneru Ramya **and Vinutha U. Muktamath, 2022,** Cyberchondriasis : An emerging challenge in digital era, ICAAAS–An Edited Book (Volume-1), *Published by*: Astha Foundation, Meerut (U.P.) India, PP:116-119, **ISBN : 978-81-958010-0-8**
4. Vinutha U. Muktamath, **2022,** Recent Advances in Home Science, Vol 3**,** Role of Breast Milk on Brain Development, Integrated Publications, New Delhi, **ISBN-978-93-93502-03-2.**
5. **V. Muktamath, U**., R. Hegde, P., Koneru, R., & Lakkashetti, R. (2023). Breastfeeding Practices and Infant Development Outcomes. IntechOpen. doi: 10.5772/intechopen.111867

Details of Projects completed during the last 5 years

S. No	Title	Cost in Lakh	Start Date	End Date	Role as PI/Co- PI	Agency
1	Establishment of day care centre for elderly & young children	10.0	August 2015	March 2020	PI-2019-2020 Worked as Co-PI- Up to 2018 March	Government of Karnataka
2	Post Natal Care of the Mother and New Born	1.5	June 2017	March 2018	PI	Staff Research Project, University of Agricultural Science, Dharwad

Total Projects handled-5

Awards Received

1. Kannada Rajyotsava Prashasti by Karnataka Journalists cultural guild & Karnataka cultural academy on 28.12.16
2. SERS Best Thesis Award-2018 in1st Indo-Asean Conference on Innovative Approaches in Applied Sciences and Technology held at Nong Lam University, Ho Chi Minh city , Vietnam held from 13-17 th June 2018

3. Certificate of Appreciation for the oral presentation of the paper titled " Breast Feeding Practice: Effect on infant Morbidity" in XXXI Biennial Conference of Home-Science Association of India held at Udaipur form1-3 February, 2018
4. **First position in oral presentation on** "Postnatal care of mothers: Knowledge, nutritional status and utilization of postnatal care services" at National Seminar on Role of Community science Education in Rural Development held from 21-22 June 2019 at Panthnagar
5. Dr.APJ Abdul Kalam Life Time Achievement National award by International Institute for social & economic reforms(R), Bengaluru on August 2019
6. International excellent Author Award by Saksham Society, Jaipur in International Women's Summit & Conference on " Women Global Leaders: Vision of Tomorrow at International conference held on 30.0702021
7. First best **oral presentation Paper-** Impact of Health Care & Nutrition intervention for rural elderly of Dharwad in International conference on Scientific & Educational Research Society Meerut, India and Graphic Era Hill University Dehradun,U.K., India from 9-10 th Oct'2021
8. First best **poster presentation-**Community Science disciplin0e in Farm science Universities & NEP-2020-Challenges and Opportunities at NAHEP, National conference on NEP-2020 for Comprehensive transformation in Agricultural Education-Initiative challenges and a way forward (15-16 March 2022) University of Agricultural Sciences, Raichur, Karnataka
9. Most impactful article of the month for the article "Importance of Child care quality during early years" published in International magazine Vigyan Varta, An international e-magazine for Science enthusiasts(E-ISSN: 2582-9467) on February, 2022
10. Best oral Presentation for the paper on 'Differential effects of breastfeeding practices on Infant's cognitive development" 34 th Biennial National Conference of Home Science Association of India, HSAI, at ST. Teresa'a College , Kochi, Kerala, India from 15 th – 17 th December 2022.
11. Best oral Presentation for the paper on " Nutritional and Mental Health Status of Postpartum Mothers: Predictors & Risk Factors" at VII International Conference on "Innovative Approaches in Basic & Applied Sciences for Societal Development" at the College of Basic Sciences & Humanities, GBPUAT, Pantnagar, U.S. Nagar, Uttarakhand,

India organised by SERS, Meerut and GBPUAT, Pantnagar on 24-25 th March 2023.

3. Dr. Prema B Patil, Professor and Head, Department of Human Development, College of Community Science, UAS, Dharwad

The scientist has expertise in gerontology. She has carried out her Msc. Research work on depression among elderly. Her PhD research was on the psycho-social scenario of the aged. She has worked in a project entitled "Increasing intergenerational solidarity and coping with aging" funded by ICAR, New Delhi.The students guided by her have worked on quality of life, loneliness and nutritional status among institutionalized and non-institutionalized elderly and effects of migration of adult children on quality of life and psychological well-being of elderly. She has implemented a DST funded project on Integrated services for senior citizens which was carried out in urban area of Bijapur taluk, Karnataka. The results revealed a positive and significant relationship at 1per cent level between family status and financial status, health status, family solidarity and social status of the elderly. The health status of the elderly had a positive and significant relationship with family status, financial status and family solidarity. Further family solidarity had a positive and significant relationship with family status, financial status, health status and social status of the elderly. Finally, social status of the elderly had a positive and significant relationship with family status, financial status and family solidarity.

Publications

1. Gaonkar,V. and Patil, P.B.(2007) Slow learners: Identification & strategies to accelerate science learning. Indian Psychological review, 68(4):207-212.
2. Patil P.B.(2009) Identification and development of instructional strategies to accelerate mathematics learning among slow learners. Asian Journal of Home Science, 3 (2):108-110.
3. Patil,P.B. (2016) Impact of stimulatory play materials on psycho-motor development of infants. Indian Journal of Applied Research, 6(9):5-6
4. Massar, D. and Patil, P. B. (2019) Factors influencing parenting practices of adolescent parents. Indian journal of pure and applied biosciences, 7(6): 91-99.
5. Massar, D., Patil, P.B. and Pujar, L. (2020) Socio demographic factors as predictors of adolescence resilience: Locale wise study. International journal of current microbiology and applied sciences, 9 (9): 91-99
6. Massar, D., and Patil, P.B. (2020) Effects of parenting practices and marital satisfaction on adolescent behavior problems. International journal of current microbiology and applied sciences, 9 (3): 109-120.

7. Patil, P.B., Biradar, S.A., and Jadhav, S.R. (2020) Nutri-garden: A low cost sustainable approach to improve nutritional security of farm families. International journal of current microbiology and applied sciences, 9 (11): 3662-3666.
8. Massar, D., Patil, P.B., Pujar, L. and Muktamath, V., 2021, Parenting practices as predictors of adolescents externalizing & internalizing problems. Journal of community mobilization and sustainable development, 16 (1): 1-6.

Author 4

1. **Name** : Ms. Sunkara.Hanna

2. **Present Position** : Teaching Assistant

: Avinashilingam Institute of Home Science and higher education for Women, Coimbatore,Tamil Nadu,641038

3. **Residential Address** : 2-9/1,Ambedkar colony, I[st] lane, Martur, Prakasam

District,Andhra Pradesh, 523301.

4. **Contact Numbers** : 9121987827, 9008453357

5. **Email** : hannasunkara999@gmail.com &hannasunkara517@gmail.com

6. Date of Birth : 22-04-1999

7. Nationality : Indian

8. Religion : Hindu- Christian

9. Caste : SC

1. Academic Qualifications

Examination/ Degree Passed	Specilization	Insitutuion/Board/ University	Year of Passing	Marks
B.H.Sc (Home Science)	Human Development and Family Studies	University of Agricutural Sciences,Dharwad	2023	86.7
B.H.Sc (Community Science)	Human Development and Family Studies	Acharya N G Ranaga Agricultural university,Guntur	2021	76.7
Intermediate	Bi.Pc	Vignana Bharathi Junior Colleges,Chirala	2017	96.1
S.S.L.C	-	M.S.N Z.P High School Martur	2015	87.0

2. Additional Qualifications

Cracked and qualified UGC NET JRF held on 22-04-2024 in the subject Home Science.

Qualified in ICAR SRF AICE 2024 Ph.D in Human Development with All India 12th rank and Category 5th rank.

Qualified in ICAR JRF AIEEE 2021 M.Sc in the Subject Home Science with All India 16th rank and Category 2nd rank.

FIELD OF SPECILIZATION IN TEACHING

Marriage and Family relations, Basics of Child Development, Perspectives in Marriage and Parenthood, Personality Traits and Development.

3. Research Experience

Title of M.Sc Thesis: Health Status and Memory function among Elderly.

Institution	Postion Held	Year
Department of Human Development,AVINUTY,coimbatore	Teaching Assistant	01.07.2024- Till date
Department of Human Development,AVINUTY,coimbatore	Remedial coaching in-charge	01-08-2024-Till date
Dakshniya Institute for the mentally Handicapped, Guntur	Assessementor in Special educator	23-11-2020 -27-01-2021

4. Research Publications

	National	International	Total
Research papers published		02	02
Books Published		01	01

5. Other Academic Related Publications

Popular articles	5	-		05

6. Research Papers Presented in Conferences / Seminars

Research Papers presented	National	International	Total
Poster papers presented	01	01`	2
Oral papers presented	-	01	1
Total			3

7. Participated in Seminars / Conferences / Workshops / Training Programmes

Training Programme -02

Workshop-01

Research Articles	
2024	Sunkara.Hanna and Lata Pujar *et.al* .,Education and cognitive function among older adults in Uttara Karnataka ,International Journal of Applied Home Science,Vol 11Issue 1&2(2024)Page 25-28(ISSN:2394-1413)
2024	Sunkara.Hanna and Lata Pujar *et.al* .,Prevalance of Health problems among elderly in rural and urban areas of Dharwad, Karnataka ,International Journal of Current Microbiology and Applied Sciences ,Vol 13(03) (2024) ISSN:2319-7692(print) ISSN:2319-7709(online)
Books Published	
2024	Hanna.Sunkara,Cognitve function among elderly, LAMBERT Academic Publication, ISBN 978-620-7-48316-7)
Popular Articles	
2024	Sunkara Hanna and Lata pujar ,Changes in Cognitive Function in Human Ageing, Agritech Today Magazine, Article ID: 240203028,June vol 2,Issue 3,June
	Sunkara Hanna , Lata pujar and Divya R regatipally,Benifits of Kangroo Mother Care for low birth and pre term infants , Agritech Today Magazine, Article ID: 240205044, August vol 2,Issue 5.
	Divya R regatipally and Sunkara Hanna, Drugery reducing technologies for major crops in Andhra Pradesh, Agritech Today Magazine, Article ID: 240206033, September, vol 2,Issue 6.
	Sunkara Hanna and Divya R regatipally, Application of AI in agriculture , Agritech Today Magazine, Article ID: 240207044, october vol 2,Issue7.

Poster presentation		
2023	6 th Post graduate Research Conference in University of agricultural sciences,Dharwad,29-30 th July, Cognitve function among rural and urban elderly	
2024	35th Bienneal Conference Of Home Science Association Of India (HASI) 2024, Effect of nutrition on cognitive function among elderly .Central Agricutural university,Umiam,Tura,Meghalaya,17-19 th January.	
Oral Presentation		
2023	8 th International Conference in Hybrid Mode on Global research initiatives for sustainable agriculture and allied sciences, Factors Influencing cognitive function amomg elderly, University of agricultural sciences,Raichur,18-20 December.	
Workshop		

2024	Industry Workshop on product security spotlight by Computer society of India,Kolkata Chapter, September 21	
Training programme		
2022	Training programme on Enterprenurship Development, organised by SC-SP,ICAR, Dharwad, 24 & 25 March 2022	
2024	Training Programme - Higher education and society on 12&13 th August at AvinashilingamInsitute of home science and higher education for women,Coimbatore.	

5. **Madhumati. K. Mantur**

D/0 KadappaMantur

Near vaibhav talkies, Banahatti 587311

Ph. No: 8073931125

e-mail- mkmantur287@gmail.com

Skills

- Teaching
- Training and extension
- Research and development
- Supervision/management
- Microsoft office
- Microsoft word
- Microsoft PowerPoint
- Microsoft excel
- Browsing Internet

Experience

- College of Community Science, ANGRAU, Guntur Worked as Teaching Associate, From 11/11/2019 to 30/08/2023
- Lecture in MurajiDesayi Residential College, Maragur, Karnataka 01.08.2017 to 04.04.2019

Education

- MHSc (Human Development and Family Studies) 2017 passed University of Agricultural Sciences, Dharawad
- B.Sc. (Home Science), 2015 passed, University of Agricultural Sciences, Dharawad
- PUC science 2012 passed, Govt. PU College, Jamakhandi

- SSLC/10th 2009 passed, Navodaya Model Girls Residential School, Hebbal
- National /State level tests NET and KSET 2017 passed

Award

- International excellent education award from Saksham Society, Jaipur

April, 2021

Activities during degree

- Worked in Rural Work Experience Programs
- Given training program to farmers, farm women's and youths regarding use of new technologies in the field of agriculture and home science, food and nutrition, govt schemes for funding Agriculture, healthmanagement.
- Completed hands on training program on broadcasting and demonstration of food habits, nutrition, diet Chart, health issues development information in KrishiBanuli Kendra and KrishiVijnana Kendra, Dharwad.

Research Articles/abstract Published

- Sibling relationship among twins and singletones, International Journal of Pure And Applied Bioscience, 6(5):266-271
- Impact of personal, parental and familial characteristics on sibling relationship of twins, Advances in Research, 16(6): 1-12
- Examine the socio-emotional behavior among monozygotic and dizygotic twins,Journal of Pharmacognosy and Photochemistry, 8(4):2080-2084
- Influence of demographic characteristics on sibling relationship of singletons, The Pharma Innovation Journal, 8(8):69-75
- Similarities ,differences, favorities and leisure time activities among co-twins and singletons: Acomparative study, The Pharma Innovation Journal,SP-10(5):533-536
- Impact of Covid-19 lockdown on family relationships The Pharma Innovation Journal, 10(5):1584-1587
- Socio- emotional behavior and sibling relationship among twins and singletons, thesis abstracts, vol-2, p-100 (Research Abstract)
- Educating Farm Families and Field Functionaries by using Developed Mobile App AALAMBANA (Abstract)
- Occupational stress and organizational and gender: Acomparative study (Abstract)

Popular article published

- Popular article published on "Ways to increase the level underground water through rain water harvesting", Valimiki Karnataka, Sep-2019

Practical Manuals Developed

- Family Counselling and Welfare
- Educational Psychology and Early Childhood Education
- Marriage and Family Dynamics
- Life span development – I
- Life span development – II
- Methods and Materials for Teaching Young Children
- Early childhood Education and Management
- Paper presented on "Sibling relationship quality among identical and non-identical twins' at 33 Biannival Conference of Home Science Association inIndia organized by Avinash Lingam Instittue of Home Science and Higher Education ForWomen.
- Poster presented on "Sibling Relationship Among Twins" at Post Graduate Research Conference (poster presentation) organized by University of agricultural sciences, Dharwad

Teaching activities

- Handled UG and PG courses (theory & practical)
- Preparation and compilation of content of department course materials
- Setting of semester final question papers
- Evaluation of answer scripts
- Organizing guest lectures and observational visits
- Assisting PG students in writing thesis

Extension activities

- Training lactating mothers on importance of Breastfeeding at Jonnalagadda, Anganawadi Center.
- One day training programme on Awareness programme on Disha ACT and Disha App

Radio talks

- Women empowerment
- Health is wealth
- Children are wealth of nation
- Importance of home work for children

- Health depends on food habits
- Pulses for better health
- Eat carrot and away from disease
- Importance of sesame